The Guide to
Colorado
Mammals

The Guide to
Colorado
Mammals

Mary Taylor Young

FULCRUM
GOLDEN, COLORADO

Library of Congress Cataloging-in-Publication Data
Young, Mary Taylor, 1955-
  The guide to Colorado mammals / Mary Taylor Young.
    p. cm.
  Includes bibliographical references and index.
  ISBN 978-1-55591-583-4 (pbk.)
1. Mammals--Colorado. 2. Mammals--Colorado--Identification. I. Title.
  QL719.C6Y68 2012
  599.09788--dc23
                            2011045808

Printed in Korea
0 9 8 7 6 5 4 3 2 1

Design by Jack Lenzo
Cover image of Rocky Mountain bighorn sheep (Ovis canadensis canadensis) © Bob Gress

Fulcrum Publishing
4690 Table Mountain Dr., Ste. 100
Golden, CO 80403
800-992-2908 • 303-277-1623
www.fulcrumbooks.com

Contents

Acknowledgments

Thanks to Haley Berry, Faith Marcovecchio, Jack Lenzo, and all the folks at Fulcrum Publishing for their commitment to making this another great guide to Colorado's outstanding natural heritage. Many thanks to Dave Armstrong, *Professorus extraordinarius* (and professor emeritus of ecology and evolutionary biology and environmental studies at the University of Colorado at Boulder), for his peer review of the manuscript.

Author's Note

The path of my life as a zoologist and nature writer was set during my childhood, when I was fortunate enough to be surrounded by the beauty and wild creatures of the Rocky Mountains.

I spent my summers at my grandparents' cabin in Estes Park, literally next door to Rocky Mountain National Park. We had a view of Longs Peak across the valley and the giant rock beaver who, my granddad told me, was forever climbing toward the summit of the mountain. We awoke to mule deer peering in the windows and hummingbirds buzzing around the red-trimmed feeders; we spent the days chasing chipmunks across the boulders of Deer Mountain and the nights listening to coyotes howling in the dark.

My fascination with wildlife, and how each creature is adapted in its own way to its corner of nature, began in those days. I hope this book will help others discover and marvel at the diverse wild animals that inhabit our beautiful state and that, as a result, they will support the conservation of these animals and their habitats. Like that beaver relentlessly climbing Longs Peak, preserving our natural heritage is an ongoing job.

Introduction

What Is a Mammal?

Warm and *furry*. Those are probably the first two words most of us think of in connection with mammals. And they are two of the characteristics that define the animals grouped together in the class Mammalia.

The first mammals appeared about 200 million years ago. Small, shrewlike animals moving furtively around a world dominated by dinosaurs, they survived the environmental cataclysm that followed the asteroid impact that did in the dinosaurs. From that humble beginning, mammals diversified and dispersed globally. Today there are about 4,000 species of mammals worldwide, in habitats as diverse as the Antarctic Ocean and the rainforests of Borneo.

Characteristics of Mammals

Mammals are warm-blooded animals that generate their own internal heat through the metabolic processes of their bodies, and regulate it with an internal "thermostat." Providing their own heat source allows mammals to be broadly distributed, inhabiting cold climates that are inhospitable to cold-blooded animals such as frogs and snakes.

Mammals are also covered by a protective coat of hair or fur in place of the scales of fish and reptiles and the feathers of birds. All mammals, even the cetaceans (whales and dolphins), grow at least a few hairs somewhere on their body at some point in their development.

Nearly all mammals give *live birth*. The young of fish, amphibians, reptiles, and birds hatch from eggs. (The monotremes, a group of primitive mammals that includes the duck-billed platypus, lay leathery eggs.)

The word *mammal* gives us a clue to another major characteristic of this animal group: they have mammae, or milk-producing glands. Mammals nurse their young on milk produced by the mother's body. This adaptation for extra care of the young ensures a higher survival rate. Interaction with a parent also offers an opportunity for young mammals to learn strategies and information beyond what is transferred to them genetically.

Like birds, mammals have a *four-chambered heart*, with separate systems for carrying oxygen-rich blood to the tissues and sending oxygen-depleted (but carbon dioxide–laden) blood through the heart to be reoxygenated by the lungs. This efficient means of transporting heat, oxygen, and nutrients throughout the body allows mammals to have a high metabolic rate and keep active in spite of the environmental temperature.

Mammals have *two sets of teeth*. Young mammals develop an initial set of teeth, commonly called milk or baby teeth. These are eventually replaced by a permanent set of adult teeth.

Several other anatomical features are unique to mammals. They have *three bones in the middle ear*. Each half of the *lower jaw consists of a single bone*, the mandible.

Mammals generally have well-developed sight, hearing, and senses of smell, taste, and touch. Some have reduced one sense in favor of another. Moles, whose eyes are covered with skin, have given up vision but have highly sensitive, tactile whiskers for feeling their way through their underground world.

How Mammals Are Related

Biologists classify animals in a cascading taxonomic system of categories, from general to specific, that every beginning biology student must memorize—kingdom, phylum, class, order, family, genus, species. (These categories may be broken down further into subfamily, subspecies, and so forth.) Thus, a black-tailed prairie dog would be categorized in the following way:

Kingdom Animalia (animals rather than plants)
Phylum Chordata (animals with a backbone)
Class Mammalia (mammals, as opposed to fish, amphibians, reptiles, or birds)
Order Rodentia (rodents)
Family Sciuridae (squirrels)
Genus *Cynomys* (prairie dogs)
Species *ludovicianus* (black-tailed prairie dog)

There are approximately 4,260 species of mammals in the world. Colorado is home to about 130 mammal species. Species reported only rarely in the state or based on questionable identification are not included in the species accounts, but a list of mammals with limited Colorado range is included on page 261 in the appendixes.

Where Mammals Live

The vast elevation differences in Colorado—from 3,315 feet along the Arikaree River near the Kansas border to 14,433 feet on the summit of Mount Elbert—mean our state has a great range of different habitats and plant communities where mammals make their homes. Some mammals, such as the meadow jumping mouse, are specialists limited to specific habitats. Others, like the coyote, are generalists that move among various habitats, finding what they need to survive.

Habitats Mentioned in This Guide

Grasslands are open areas in which grasses are the dominant vegetation. Forbs (flowering nonwoody plants) and shrubs also grow in grasslands, but trees are largely absent except near water (surface or ground). The Eastern Plains, much of which is now cropland, were once largely prairie grassland. Large mountain valleys such as South Park are also grassland. The term *meadow* is used for grassy areas that are not extensive and may be found within a forest.

Riparian areas are forest and shrub communities near water and along watercourses. Cottonwood, willow, birch, and alder are the main woody plants. Riparian communities are found across the state at all elevations, including within urban and suburban areas. Blue spruce grows in mountain riparian habitat.

Wetlands are areas of shallow standing or slow-moving water where water-loving plants like sedges, rushes, and cattails grow. There are wetlands at all elevations statewide.

Shrublands are areas where shrubs are the dominant plants. There may be some small trees and open areas of grasses and forbs. Montane shrublands are found in the foothills of the Eastern Slope, with shrubs such as mountain mahogany, chokecherry, and serviceberry. Semidesert shrublands are found at lower elevations of the Western Slope. Big sagebrush and desert shrubs such as greasewood and saltbush are the main woody plants.

Piñon-juniper woodlands are found throughout southern Colorado. Piñon pine growing with Utah, one-seed, or Rocky Mountain juniper (also called western red cedar) make up what is called the "pygmy forest" because the trees only grow 20 to 40 feet tall.

Mountain coniferous forests include lower elevation montane forests of ponderosa pine, Douglas-fir, and lodgepole pine, and, in southern Colorado, white fir. This category also includes dense subalpine forests of subalpine fir and Engelmann spruce, as well as limber and bristlecone pine. Blue spruce grows in moister areas.

Aspen forests are mountain forests made of nearly uniform stands of aspen.

Timberline is the highest level of tree growth below the alpine tundra. It is the edge, or ecotone, where the subalpine forest meets the tundra.

Alpine tundra is the highest habitat in Colorado. This austere land has very low-growing forbs and grasses and some shrubs, but no trees.

Agricultural land includes cultivated fields, rangeland, areas along irrigation ditches and unused fields, and the edges between them.

Urban and suburban habitats include open spaces, parks,

cemeteries, wooded stream corridors, and other land around towns, cities, and suburban neighborhoods.

How to Watch Mammals

Get out there! Most mammals are secretive. Unlike birds, they generally live their lives hidden from sight. Animal sightings are often serendipitous, so the more time you spend outdoors in their habitat, the better chance you have of seeing wildlife.

Time your outings for early morning and evening. Most mammals are nocturnal, but you have a chance of glimpsing them as they emerge in the evening or before they retire at dawn. Many diurnal animals are most active early and late in the day and rest under cover during the middle of the day.

Use binoculars, spotting scopes, and long camera lenses for close-up views. Many species will not allow you to approach closely, and good optics let you have a look at those shy creatures. They also allow you to keep a good distance between you and the animals, for the safety of both.

Move slowly and quietly to avoid disturbing wildlife. Many animals interpret noise and quick movements as danger and may flee. This can stress the animals, cause them to leave their young unprotected, and deny you the pleasure of observing them.

Wait for animals to find you by locating an area of likely animal activity and sitting quietly. Once you quiet down, animals may reemerge or pass through, giving you a chance to observe their natural behavior.

Wear earth-toned clothes, like gray, khaki, and olive green. They make you less obvious and more a part of the environment. Most mammals do not see color well, but they are very good at seeing contrasts. Except in snow, white stands out like a giant warning flag.

Stay in your car and use it as a viewing blind. The large, blocky shape of a vehicle is not as threatening to an animal as the shape of a human. Wildlife may even gradually move near and around your vehicle, giving you an unparalleled look at them.

Don't feed wildlife. Human food is unhealthy for wild animals and can cause them to overcome their fear of humans and change their behavior in ways that can be harmful to them, to you, and to the people who come after you.

Breeding season offers the chance to see, and hear, interesting behaviors. The animals are preoccupied with courtship and mating and may be less wary and therefore easier to find and see.

Learn more about wildlife so you know where to find them, what species to expect, and how to understand their behavior. Use field guides, websites, and other resources. You might want to make note of your sightings, and what you observe, in a nature journal.

Mammal-Watching Ethics and Etiquette

Respect the animals and don't disturb them, their dens, their young, or their habitat. You are entering their "home," so conduct yourself as a guest.

Don't approach any closer than an animal feels comfortable. If it alters its behavior, stops feeding, or otherwise seems agitated, back off.

Leave baby animals where you find them. Many adult animals rely upon camouflage to protect their young when they are away feeding or are frightened off. The parent is likely close by waiting for you to leave before coming back to care for its baby. If you are certain the parent is dead or has abandoned the young, contact a licensed wildlife rehabilitator. You can find information on the Colorado Parks and Wildlife website, http://wildlife.state.co.us.

Leave your pets at home or in the car. Dogs and wildlife of any kind don't mix. Even the best trained dog is likely to chase animals, from chipmunks to elk, and may dig up a burrow or den. Harassing wildlife is also illegal under some federal and state laws.

Never chase, herd, or flush animals, or make deliberate noise in an attempt to get a better look at one. Don't alter habitat by removing rocks, branches, or natural features; excavating a burrow or den; or destroying a nest.

It is illegal to kill, possess, sell, or trade native mammals or their body parts without the appropriate license. Consult the Colorado Parks and Wildlife website for more information.

Be considerate of landowners and always ask permission before entering private land to view wildlife. Never cross No Trespassing signs without permission. On public land, observe all rules and regulations and tread lightly, staying on trails and roads. In a self-serve, fee-pay area, pay your fee instead of sneaking in. These revenues come back to all of us by supporting preservation of habitat and public-use areas.

Remember that even just watching animals has an impact. Intrusion into their living space can expose them to predation, keep them from feeding or other essential activities, or cause them to leave their young exposed to predation or the elements. No photo or viewing opportunity is worth harassing or stressing wildlife. In appreciating and watching them, we have a responsibility to protect and preserve the animals that share our state.

How to Use This Book

This guide is meant as a general introduction to Colorado mammals for casual naturalists, outdoor recreationists, families, Colorado vacationers, and anyone desiring a general overview and identification guide. Further detail on the life histories of these species can be found in more scholarly works, including *Mammals of Colorado*, published by the Denver Museum of Nature & Science. Certain identification of many species requires close study of anatomy, especially teeth or skeletal features, which is beyond the scope of this book. The intent of this guide is to help in field identification of animals encountered in the wild but not necessarily held in hand or dissected.

Particular emphasis is given to interesting behaviors, especially those observable by wildlife viewers in the field. It is the author's intent that readers enjoy reading and learning about animals they might encounter as well as those they may never see. Colorado offers unsurpassed scenery, but it is the abundance of fascinating animals, with their infinite adaptations for life in the state's many habitats, that brings energy and vibrancy to our landscape.

Species Accounts

A species account for each animal includes a description and life history information. Accounts are organized taxonomically, beginning with species considered more primitive—the marsupials and shrews—and moving to the most advanced, the hoofed animals.

Common name is the accepted, nonscientific name for the species. Some profiles in this book also mention a secondary common name, if it is widely or locally familiar. The **scientific name** is the accepted, formal name of the species, expressed in two words reflecting the animal's genus and its unique species name, both italicized. Latin names follow Wilson and Reeder's *Mammal Species of the World, Third Edition*, 2005.

Size is given for adult animals. Head/body length refers to the measurement from tip of nose to base of tail. Length of tail is given separately where knowing its length may aid identification. Total length is given where the tail is inconspicuous. Height at the shoulder is given for large mammals. Where males and females vary dramatically, measurements are given for each. Measurements are given first in inches (in), feet (ft), ounces (oz), and pounds (lbs) because these are most familiar to most Americans, and then in metric—millimeters (mm), centimeters (cm), meters (m), grams (g), and kilograms (kg). Various sources were used for size data, including *A Field Guide to Mammals of North America* (Peterson Field Guides); *Mammals of Colorado* and *Mammals of North America* (Princeton Field Guides); *Bats of the Rocky Mountain West*; and the Colorado Parks and Wildlife website.

Habitat lists the basic habitats or general plant communities the species inhabits.

Distribution discusses where the species is found geographically in the state. (For clarity, "Front Range corridor" refers to the urban/suburban region along the Eastern Slope of the Rocky Mountains to the edge of the Eastern Plains, from north of Fort Collins to Pueblo.) A **range map** shows graphically where the species is found within Colorado.

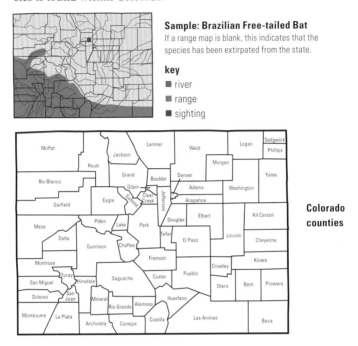

Sample: Brazilian Free-tailed Bat
If a range map is blank, this indicates that the species has been extirpated from the state.

key
■ river
■ range
■ sighting

Colorado counties

Field Notes offer details of life history, physical characteristics, behavior, and other interesting information. Technical terms mentioned in the text are defined in the glossary at the back of the book.

Legal Status is based on categories defined by Colorado Parks and Wildlife and the species' federal status, where applicable.

Sidebars offer further information or insight on different animals or animal groups, their behavior, and life history.

ORDER
DIDELPHIMORPHIA

American Marsupials: Pouched Mammals

These animals were formerly classified with kangaroos, koalas, and other Australasian marsupials within the order Marsupialia. Didelphimorphia includes just a single family, Didelphidae, the American opossums, all of which inhabit South America except one species. The lone North American pouched mammal is the Virginia opossum.

Marsupials are mammals whose young are born at a very early, almost embryonic, state of development. The young then crawl across the mother's belly into an external, fur-lined natal pouch. Here they attach to a nipple and grow and develop until they are covered with hair and large enough to begin life out in the world. Pouched mammals are a separate evolutionary branch from placental mammals (animals whose young mature within the mother's body).

New World marsupials are all correctly called *opossums*, as opposed to the often-used *possum*, which is the name of a different group of Australian marsupials. To make things more confusing, *opossum* is often pronounced with the first *o* silent.

This order includes 60-some species, ranging from the 1½ ounce mouse opossum to the up-to-15-pound Virginia opossum. Most are tree-dwellers, though the yapok, or water opossum, is aquatic, with webbed hind feet and a reversed pouch to prevent water from entering when the animal is swimming.

Family Didelphidae—Opossums

The family Didelphidae is restricted to the New World, with one species in North America and the rest ranging through Central and South America. This is the oldest known marsupial family and the most generalized. These animals have five toes on each foot and walk on the soles of the feet (plantigrade). They have an opposable toe on the hind foot that aids climbing and is probably a holdover from tree-dwelling ancestors. The word *opossum* comes from the Algonquian word *wapathemwa*, meaning "long-nosed, ugly-faced animal that is good to eat."

Animal Sign: Five-toed hind track with the "big toe" widely spread from the other four toes and to the side or backward; five-toed front track with the toes widespread and handlike.

Virginia Opossum
Didelphis virginiana

Field ID: This housecat-sized animal is grayish with a pale face, pink nose, and round, naked ears. The long, scaly, nearly hairless tail, which resembles a rat's, is dark at the base but lighter toward the tip.

Size: Head/body length 15–20 inches (38–51 cm), tail 9–15 inches (23–38 cm), weight 4½–15 lbs (2–7 kg).

Habitat: Riparian areas near agricultural land.

Distribution: Along the South Platte, Republican, and Arikaree rivers in eastern Colorado, and around Grand Junction in western Colorado (from introductions in the mid-twentieth century).

Field Notes: The opossum is the only marsupial native to North America. A secretive, nocturnal animal, it hides during the day in thickets, burrows, wood or rock piles, and hollow trees. Opossums may be more abundant in some areas than commonly thought because of their secretive habits. Opossums eat carrion, insects, fruit, and the occasional small animal. They are known for "playing possum"—rolling over, closing their eyes, and lolling out their tongue to feign death when threatened. Their ratlike, prehensile tail can grasp a branch or carry nesting material. Opossums may have up to 25 young in a litter, though only 8 or 9 survive to emerge from the natal pouch three months later, often because the mother has more young than nipples for them to attach to. The female carries her young, which resemble black-eared rats, on her back with their tails wrapped around hers. The inside toe on the hind foot is opposable, like a human thumb, enabling the opossum to securely grasp branches as it climbs. Opossums are active year-round but reduce their activity during cold weather.

Legal Status: Furbearer.

ORDER
SORICOMORPHA

Shrews and Moles: Insect-eaters

The soricomorphs, or shrew-shaped animals, include shrews and moles—small mammals that may somewhat resemble the first mammals that emerged in the Triassic Period when the Earth was dominated by dinosaurs.

These animals were once classified taxonomically in an order known as Insectivora (insect-eaters). Their pointed teeth are adapted for feeding on insects and other invertebrates, and they have canine teeth (unlike rodents). They have five toes on each foot.

Vision is not strongly developed in these animals—the mole's eyes are covered with skin—but scent is important in locating prey. Tactile whiskers help them orient as they move along trails beneath vegetation and in subterranean burrows. Shrews use echolocation for orientation, emitting a high-pitched twittering—a useful technique for an animal that often operates with little light.

Family Soricidae—Shrews

Shrews are very small and superficially resemble mice. They are often mistaken for rodents, but shrews are carnivorous and have long, very pointed noses and tiny eyes. They have canine teeth, but their incisors are unlike the long, chiseling teeth of rodents. With the exception of the relatively large water shrew, species of shrews can be very difficult to distinguish from each other.

The shrew's metabolism is a raging furnace. The heart beats 750 times a minute and the animal breathes 168 breaths per minute. A shrew patrols runways and burrows in a constant search for food, with brief pauses to sleep and digest. It consumes several times its body weight in food every day. Shrews prey on insects, worms, and the young of other small mammals. Many shrews produce a toxin in their saliva that paralyzes prey, allowing them to kill animals much larger than themselves.

Animal Sign: Tiny (1/4 inch wide) track with five toes on all feet, showing a tail drag. Unlikely to leave a track in anything except snow due to its light weight.

Masked Shrew
Sorex cinereus

Field ID: This medium-sized shrew is grayish brown with paler underparts, a pointed nose, tiny eyes, and a long tail that is slightly darker on the top.

Size: Head/body length 2–2½ inches (51–64 mm), tail 1¼–2 inches (32–51 mm), weight ¹/₁₀–¹/₅ oz (3–6 g).

Habitat: Wet meadows and willow thickets of montane and subalpine areas of the mountains; wetlands along the mountain front.

Distribution: Foothills and mountains of central and western Colorado and along the Front Range corridor, from about 5,000 feet to 11,000 feet.

Field Notes: The masked shrew is often the most common shrew in mountain habitats and is difficult to distinguish from the montane shrew. Its range, at the western edge of the Great Plains, is broader than that of the montane shrew and it seems to be more common on the Eastern Slope. It remains active throughout the year, operating above-ground but beneath leaf litter and other cover, under the snow, and in burrows it builds or in the burrows of other animals. It searches constantly, day and night, for worms, insects, grubs, and other invertebrates, alternating bouts of activity with rest periods. Because of a very high metabolism—its heart beats 1200 times per minute!—the masked shrew eats more than its own weight in food each day. In summer, females bear multiple litters of four to ten young, which are mature in four weeks.

Legal Status: Nongame.

Pygmy Shrew
Sorex hoyi

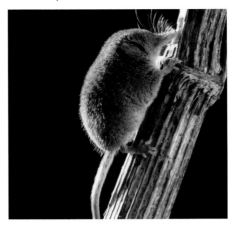

Field ID: This tiny shrew is dark brown with a pale underside; long, pointed nose; and tiny, bead-like eyes. The tail is slightly bicolored and fairly short.

Size: Head/body length 2–2½ inches (51–64 mm), tail 1–1⅖ inches (25–36 mm), weight ¹⁄₁₀–¹⁄₇ oz (3–4 g).

Habitat: Mountain forests, wet meadows, willow thickets, forest-meadow edges.

Distribution: Central northern Colorado above about 9,600 feet.

Field Notes: The pygmy shrew is in the running, along with the dwarf shrew, as the smallest mammal in the world. It weighs no more than a dime. It wasn't found in Colorado until 1961 and is still not known in many locations. It may be more widespread than records indicate due to the difficulty of finding and collecting specimens. The pygmy shrew lives in a variety of habitats, including subalpine forests and meadows, and eats small insects, worms, other invertebrates, and carrion. Like all shrews, it's active constantly, year-round. It spends its days and nights alternating between bouts of frantic foraging and resting. It builds its nests and runways beneath leaf litter and fallen logs. Pygmy shrews breed in the warm months, the females producing a single litter a year of up to eight young.

Legal Status: Nongame.

Merriam's Shrew
Sorex merriami

A photo of this species is not available. See the photo of the masked shrew on page 18 for a similar species.

Field ID: A medium-sized, pale gray to grayish brown shrew with white underparts and white feet, pointed nose, and inconspicuous ears. The nearly hairless tail is distinctly dark on top and pale underneath.

Size: Head/body length 2¼–2½ inches (57–64 mm), tail 1¼–1⅝ inches (32–41 mm), weight ⅛–¼ oz (4–7 g).

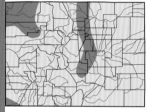

Habitat: Montane, sagebrush, and semi-desert shrublands; grasslands; piñon-juniper woodlands; montane and sub-alpine forests.

Distribution: Foothills and lower mountains of the central and western parts of the state between about 4,500 and 9,600 feet.

Field Notes: This shrew may be more common and widespread in our state than previously thought. It is a western shrew and while it hasn't been greatly studied, it seems to live in drier habitats, such as foothills shrublands and sagebrush, than other shrew species. It patrols the runways of other animals such as voles, feeding on ground-dwelling invertebrates, including caterpillars, crickets, beetles, and spiders. Evidence from bones found in owl pellets indicates this shrew is probably active aboveground during the night. Among western shrew species, Merriam's shrew has the strongest bite, allowing it to eat large, hard-shelled insects like beetles. It builds a nest hidden beneath logs. Females have one to two litters a year of four to seven young.

Legal Status: Nongame.

Montane Shrew
Sorex monticolus

A photo of this species is not available. See the photo of the masked shrew on page 18 for a similar species.

Field ID: A stout, dark brown shrew with a fairly long tail that is dark on top and light underneath. It is slightly paler on its underside.

Size: Head/body length 2–3 inches (51–76 mm), tail 1⅓–2 inches (35–51 mm), weight ⅛–¼ oz (4–7 g).

Habitat: Aspen stands, willow thickets, mountain wetlands, mountain meadows, forest edges.

Distribution: Foothills and mountains of central and western Colorado from about 5,300 feet to 11,500 feet.

Field Notes: As shrews go, the montane shrew is fairly large, though only a little larger than the masked and Merriam's shrews, and notably smaller than the American water shrew. It likes moist mountain habitats and in some places is common around beaver ponds and other flooded, marshy, or boggy places. Montane shrews inhabit areas with downed logs and other cover where they remain mostly out of sight. They are active day and night, year-round, moving around beneath the snow in winter. The montane shrew is hard to distinguish from the masked shrew, which occupies the same habitat but is slightly smaller and more grayish. The montane shrew is probably more common than the masked shrew on the Western Slope. Its nest is made of grass and leaves. Montane shrews breed April through August, bearing one or more litters of two to nine young.

Legal Status: Nongame.

Dwarf Shrew
Sorex nanus

Field ID: A tiny shrew with pale, grayish brown body and long tail that is faintly two-colored.

Size: Head/body length 2–2¼ inches (51–57 mm), tail 1–1¾ inches (25–44 mm), weight ¹⁄₁₀ oz (3 g).

Habitat: Rocky alpine and high-mountain slopes, mountain wetlands, forest bogs, brushy hillsides, open woodlands, coniferous forests.

Distribution: Throughout much of central and western Colorado and along the Front Range corridor from Wyoming to south of the Arkansas River from about 5,300 to 10,000 feet.

Field Notes: The dwarf shrew is in contention with the pygmy shrew as the smallest mammal in Colorado, and in the world (the lesser shrew of Europe weighs into this competition as well). The dwarf shrew is longer overall than the pygmy, due to its longer tail, but has a shorter body and weighs very slightly less. It also prefers drier habitats and has been collected in grasslands and dry hill-sides. It feeds on insects, spiders, and the carrion of mice and other vertebrates. Dwarf shrews build a nest under a log or in soft soil. Females probably have two litters of four to eight young a year. Not a lot is known about the dwarf shrew, which is difficult to collect because of its tiny size and hidden habits. It may be more common and widely distributed than presently thought.

Legal Status: Nongame.

American Water Shrew
Sorex palustris

Field ID: This large shrew is blackish or charcoal gray, with a silvery underside and a long tail. The tail is two-colored and as long as the body. The hind feet are large and have a fringe of stiff hairs along the toes and sides.

Size: Head/body length 3¼–3½ inches (83–89 mm), tail 2½–3 inches (64–76 mm), weight ½–⅔ oz (14–19 g).

Habitat: Riparian areas, streamsides, riverbanks, wetlands, along ponds and lakes.

Distribution: Throughout the mountains of central and western Colorado from about 7,000 to 10,000 feet.

Field Notes: Among Colorado shrews, the water shrew is a behemoth, weighing up to four times as much as the pygmy shrew. It is well named as it is adapted to life in and around water. It swims not only in the water but also on top of it, the stiff hairs on its feet allowing it to scamper short distances across the surface like a water strider without breaking the surface tension. Air bubbles trapped in the hairs of the shrew's velvety coat may keep the animal dry while also giving it a silvery sheen. Anglers sometimes report seeing a small, silvery mouse swimming under the water—almost certainly a water shrew. It lives along rivers, streams, ponds, lakes, and wetlands, eats mainly aquatic insects, and can capture small fish. Water shrews are sometimes caught in fish traps or on fishing lines. Their nests of dried leaves and sticks are set within cover around waterways, sometimes within beaver lodges. Between January and August, females bear multiples litters of five to seven young.

Legal Status: Nongame.

Elliot's Short-tailed Shrew
Blarina hylophaga

Field ID: This is a stout, silvery-gray to nearly black shrew with a tail that is less than one-third the length of the head and body. It has a smooth, round-headed appearance because it lacks external ears. The ear openings are hidden by fur.

Size: Head/body length 3–4 inches (76–102 mm), tail ¾–1¹⁄₅ inches (19–30 mm), weight ½–1 oz (15–30 g).

Habitat: Riparian woodlands, marshes, moist brushy areas, fields, and meadows.

Distribution: Along the Republican River in far northeastern Colorado.

Field Notes: Only a very few specimens of this shrew have been found in Colorado, in the far northeastern edge of the state. This species was formerly considered the same as the short-tailed shrew, *Blarina brevicauda*, which is broadly distributed throughout the eastern United States from the Great Plains to the Atlantic coast. Elliot's shrew builds burrows under leaf litter or dug into loose soil as much as 20 inches below ground, where it makes a nest from grass, leaves, and fur. It breeds between February and October, producing one to two litters of between three and ten young. Elliot's shrew eats insects, worms, and snails and also hunts snakes, birds, and other small mammals. A toxic venom in its saliva allows it to incapacitate prey. While other shrews have a constantly raging metabolism, Elliot's shrew lowers its metabolism at times, perhaps to survive cold temperatures. It also eats only about half its weight each day.

Legal Status: Nongame.

Least Shrew
Cryptotis parva

Field ID: This small, gray-brown to cinnamon-brown shrew has a long, pointed nose and short tail. It has pale to white undersides.

Size: Head/body length 2⅕–2½ inches (56–64 mm), tail ½–¾ inch (13–19 mm), weight ⅐–¼ oz (4–7 g).

Habitat: Grasslands, old fields, marshy areas, riparian woodlands, suburban areas around houses and structures.

Distribution: Along the South Platte River from the Front Range to the state line, north along the Front Range into Larimer County and in the Republican River drainage.

Field Notes: Mainly nocturnal, the least shrew patrols the burrows it constructs beneath leaf litter and along the ground surface, seeking worms, insects, spiders, snails, and larvae. It often also uses the runways of voles. It prefers wetter habitats, where it can root in the soil for prey, digging with its nose and shoving the dirt behind itself with front and back legs. Because of its high metabolism, the least shrew may consume its body weight in food each day. While most shrews are solitary animals, least shrews have been found in large groups of up to 30 animals. In winter, they cluster in leaf-lined nests at or below the ground surface, and have been found in beehives, where they may benefit from the hive-warming activities of the bees. Least shrews breed in spring and summer, bearing one or more litters of three to seven young. The babies grow to adult size in one month. This species' short tail and cinnamon color help distinguish it from other shrews.

Legal Status: Nongame.

Desert Shrew
Notiosorex crawfordi

Field ID: This small, slender shrew has a long, pointed snout and a tail that is less than one-third of its total length. It can be silvery to dark brownish gray, with paler undersides. It has visible external ears and tiny eyes.

Size: Head/body length 2–2¾ inches (51–70 mm), tail 1 inch (25 mm), weight ¹/₇–¹/₅ oz (4–6 g).

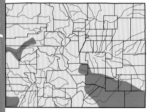

Habitat: Semidesert shrublands, riparian areas, grasslands, piñon-juniper woodlands, mixed piñon-ponderosa pine forests, rocky areas.

Distribution: In southeastern Colorado south of the Arkansas River, the far southwestern corner of the state, the far western part of the Colorado River.

Field Notes: The desert shrew is well named, living in dry habitats of the Southwest and Mexico. It is adapted to live without free water but will drink water when available. It is not discriminating in its food choice, feeding on moths, beetles, crickets, and grasshoppers, and the carrion of reptiles, birds, and mammals. Captive desert shrews would not kill and eat rodents, scorpions, or worms offered as food. They construct nests of plant fibers and hair beneath debris, plants, or small shrubs. The desert shrew also sometimes takes up residence within a woodrat nest. When food is abundant, it tolerates others of its species, unlike most shrews, which are aggressive toward other shrews. The litters of three to five young grow to adult size in 90 days.

Legal Status: Nongame.

Family Talpidae—Moles

Moles live their entire lives in burrows in the ground. Because they have no need for vision in their lightless world, their eyes are covered with skin, an adaptation to keeping dirt out of organs that are no longer needed.

The front feet of moles are enormous for the size of the animal—flat, spade-shaped, rotated outward, and equipped with very long, sharp claws for digging.

Animal Sign: Raised, low ridges of soil pushed up by the mole's tunneling; mounds of loose dirt with no apparent entrance.

Eastern Mole
Scalopus aquaticus

Field ID: A stocky-bodied, gray to brown animal with short, soft fur and short, sparsely haired tail. The snout is pointed, with a naked, flexible tip. The paws are spade-shaped. Eyes are not detectable on the outside, and there are no external ears.

Size: Head/body length 4½–6½ inches (114–165 mm), tail ¾–1½ inches (19–38 mm), weight 2¼–3½ oz (64–100 g).

Habitat: Moist, loose soil of riparian areas, lawns, golf courses, fields, grasslands.

Distribution: The northeastern corner of the state in the South Platte and Republican river drainages, and in the far southeastern corner of Baca County.

Field Notes: Moles spend their entire lives in dark galleries in the ground, feeling their way with their sensitive nose, whiskers, and paws. They live in soft, loose loamy, or sandy soils but not in hard clay or gravelly soils. These small creatures are digging machines, their front paws shaped like spades and tipped outward to aid digging. They have extremely short legs and sharp claws, since they use them more for digging than for locomotion. With no need to detect light, their eyes are vestigial and covered with thin skin. They lack external ears and the ear openings are tiny holes— an advantage for keeping out dirt. Moles build shallow burrow systems, which show on the surface as raised dirt ridges, along which they forage for worms, insects, and grubs. Their deeper burrows are for nesting and access between shallow runs. Their nests are made of dry leaves and grass. Moles breed in March and April. Females bear a single litter of three to five young, which mature in one month.

Legal Status: Nongame.

ORDER
CHIROPTERA

Bats: Winged Mammals

Bats are the only mammals capable of true, self-powered flight (flying squirrels actually glide). They have leathery wings that connect the "hands" and "arms" to the sides of the body and the hind legs. A second membrane connects the hind legs and the tail. In free-tailed bats, much of the tail projects beyond this interfemoral membrane.

A bat's wings are supported by the bones of a highly modified hand. The name of the order Chiroptera means "hand-wing." The bat's highly flexible wing allows it to make quick and intricate aerodynamic movements, which is why bat flight often looks erratic and fluttery. Constant adjustment of the shape of the wings allows the bat to stop, turn, and swoop after the maneuverable insects it feeds on.

Contrary to common perception, bats are not blind and actually see well, but they rely on other senses in the dark.

Many bats locate their prey, and avoid obstacles, using echolocation. They fly with their mouths open, emitting a rapid series of ultrasonic squeaks and interpreting the echoes that bounce back from objects. They roost by hanging upside down, secured by the very sharp claws of their hind feet, which are rotated at the ankle to create a straight line with the lower leg.

Many bats, especially those that hibernate rather than migrate in winter, have evolved the biological mechanism of delayed fertilization. Animals mate in fall, but the female retains the sperm in her body over winter and does not ovulate until she emerges from hibernation, when eggs are fertilized and gestation proceeds. Many bats give birth hanging upside down in their roosts. Others crawl around so they are head upward, hanging by their thumb claws. They catch their babies in their tail membranes as they are born.

About one-fourth of all mammal species in the world are bats, but many are small, fragile animals. Persecution of bats due to cultural fears, along with habitat loss and disease, threatens bat populations in Colorado and throughout North America. Bats are highly beneficial animals, consuming millions of pounds of insects a year, a large percentage of this being the larvae of agricultural pests. A little brown bat may capture 600 mosquitoes per hour.

Family Vespertilionidae—Evening Bats

This large family of bats includes the genus *Myotis*, which is the most diverse and widespread genus of bats in Colorado and North America generally.

Vespertilionids are sometimes called plain-nosed or evening bats. The muzzles of this family of bats are simple and unmodified, without the knobs and protrusions that give the faces of many bats a "gargoyle" look. The tail, or interfemoral, membrane is complete between the hind leg and tail. The tail extends to the edge of the membrane but does not protrude beyond it.

Animal Sign: Flight pattern (see Identifying Colorado Bats in Flight sidebar on page 50); guano (droppings) accumulated below a roost, but not specific to family or species.

California Myotis
Myotis californicus

Field ID: This small bat is usually pale yellowish brown but can be almost black. The bases of the hairs are much darker than the tips. It has large, dark ears.

Size: Total length 2¾–3¼ inches (70–83 mm), wingspan 9–10 inches (23–25 cm), weight ¹/₁₀–¹/₆ oz (3–5 g).

Habitat: Dry shrublands, piñon-juniper woodlands, mines, buildings.

Distribution: Up to about 7,500 feet in far western Colorado along the Utah state line and eastward along the Yampa and Colorado rivers; the south-western corner of the state.

Field Notes: The California and small-footed myotis bats can be difficult to distinguish from each other—both of them tiny bats with relatively long ears. This myotis crawls into narrow crevices to roost—under tree bark, in hollow trees, in caves or mines, and in buildings. It emerges soon after sunset, foraging only about 6 to 9 feet above the ground for moths, flies, spiders and other small invertebrates. It may return to roost several times in the night between foraging bouts. It also hunts over arroyos and riparian areas, stock tanks, cliffs, and open country. The California myotis breeds in the fall but the sperm is stored in the female's body. If she is in good enough shape and not too weak after winter, fertilization and implantation take place in the spring, and she usually bears a single young. The California myotis probably hibernates through winter in Colorado instead of migrating to warmer climates.

Legal Status: Nongame.

Western Small-footed Myotis
Myotis ciliolabrum

Field ID: This small bat has long, pale brown to yellowish fur and often a dark mask across the face. The tips of the hairs are shiny, giving the coat a metallic sheen. It has dark ears and wings and very small feet.

Size: Total length 3–3½ inches (76–89 mm), wingspan 8–10 inches (20–25 cm), weight $1/7$–$1/5$ oz (4–6 g).

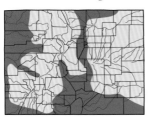

Habitat: Rocky, broken terrain; rock formations; caves; buildings.

Distribution: Far western Colorado, southern Colorado from the San Luis Valley to the southeastern corner, and northward along the Front Range corridor to about the Wyoming state line.

Field Notes: The small-footed myotis leaves its daytime roost beneath tree bark or in a crevice, mine, or building in early evening, while it is still light. It pursues its prey in a slow, fluttery flight, hunting low—within six feet of the ground—over rocks and shrubs and around cliffs and open areas. It eats ants, beetles, and flies. Hunting early, as other small bats do, may allow it to avoid being preyed on by larger bats such as the hoary bat.

The small-footed bat inhabits Colorado year-round. It hibernates singly (not in a colony with others of its species) in a cave, mine or tunnel, often sharing space with much larger species such as Townsend's big-eared or big brown bats. These bats bear a single young (occasionally twins) in summer. Females gather in nursery colonies of 10 to 20 mothers plus their babies.

Legal Status: Nongame.

Mouse Ears

The genus *Myotis* is the largest and most widespread genus of North American bats. It includes 7 of the 17 bat species covered in this guide, including the common little brown bat (also called little brown myotis).

Bats are often described as "mice with wings" because of their tiny size and large ears. No surprise, then, that the word *myotis* comes from two Greek words—*mys*, meaning "mouse," and *otos*, meaning "ear." *Myotis*, then, means "mouse-eared."

Despite their superficial resemblance in terms of size (and in some cases ears), bats and mice are vastly different animals not at all closely related. The common ancestor of the two groups probably lived in the Age of Reptiles, 100 million years ago or more.

Long-eared Myotis
Myotis evotis

Field ID: This pale brown bat has very large, black ears. The wings and tail membrane are also black.

Size: Total length 3½ inches (89–92 mm), wingspan 10–12 inches (25–30 cm), weight ⅕–¼ oz (6–7 g).

Habitat: Ponderosa pine and piñon-juniper woodlands.

Distribution: Western Colorado, San Luis Valley, and along the mountains and lower elevation foothills of the Eastern Slope between 6,000 and 9,000 feet.

Field Notes: The long-eared myotis is well named, with the longest ears of any myotis in North America. Laid forward, the ears extend beyond the muzzle. This bat emerges to forage late in the evening. It is a bat of coniferous forests, mostly ponderosa pine but also piñon-juniper. Preferring to feed along the edges of trees and over water, it hovers like a long-eared, nocturnal hummingbird, picking insects off of leaves. Flies, beetles, moths, and spiders make up much of its diet. It roosts in hollow trees, under tree bark, in caves and mines, and in buildings. Like the California myotis, the long-eared myotis breeds in fall, but fertilization and embryo implantation are delayed until spring. A female typically bears a single young. Nursing females gather in small maternity colonies that range in size from just a few to a couple dozen animals. It is unknown whether this species migrates or hibernates in Colorado.

Legal Status: Nongame.

Little Brown Bat/Little Brown Myotis
Myotis lucifugus

Field ID: A medium-sized, light to dark brown bat with paler undersides and longish, shiny fur. The ears are only medium-large.

Size: Total length 3½–4 inches (89–102 mm), wingspan 9–11 inches (23–28 cm), weight ⅓–½ oz (9–14 g).

Habitat: Riparian, suburban and urban woodlands, farm shelterbelts.

Distribution: Statewide up to 11,000 feet.

Field Notes: This is the most studied bat in North America and probably also the most common. It eats moths, beetles, flies, mosquitoes, and other flying insects. But instead of snapping up insects, it uses its wings or tail membrane to net prey from the air, then gobbles the meal from this "catcher's mitt." The little brown bat emerges at dusk, skims a nearby pond for a drink, then begins hunting. Bats captured within an hour of emergence have been found to have already consumed as much as one-fifth of their weight in insects. Returning to a night roost to rest, the little brown bat makes several additional feeding flights in a night. Also called little brown myotis, it has adapted well to life around humans, making use of urban and suburban woodlands as well as buildings and bridges. Large groups sometimes congregate in attics. Some hibernate in Colorado in caves, mines, and buildings. Mated females store sperm through winter. Eggs are fertilized once females emerge from hibernation. While many bats give birth in the head-down, roosting position, the little brown bat female reverses position and hangs by her thumb claws, catching her single baby in her tail membrane as it is born. In Colorado, nursery colonies number 100 or fewer mothers plus their young.

Legal Status: Nongame.

Fringed Myotis
Myotis thysanodes

Field ID: A yellowish to reddish brown bat with large, dark ears. The edge of the tail membrane has a fringe of stiff hairs.

Size: Total length 3–4 inches (76–102 mm), wingspan 11–13 inches (28–33 cm), weight ⅙–¼ oz (5–7 g).

Habitat: Ponderosa pine forests, desert and oakbrush shrublands.

Distribution: In far western Colorado, along the Colorado and Gunnison rivers, along the foothills of the Front Range and into southeastern Colorado below 7,500 feet

Field Notes: The fringed myotis does not seem to be a common bat in Colorado. It garners its name, and its definitive characteristic for identification, from the fringe of stiff hairs along the trailing edge of its tail membrane. It also has large ears. The fringed myotis roosts in colonies in caves, mines, and buildings. It leaves the roost site to hunt soon after sunset, continuing to be active for several hours after dark. Moving in a slow flight, it forages along the edges of shrubs and trees, hovering to glean insects off the leaves and branches. It feeds on moths, ants, bees, caddis flies, beetles, and other insects. Mated females store sperm over winter, bearing a single young in early summer. In some parts of its range, the fringed myotis gathers in large nursery colonies of hundreds of bats. It hibernates in winter in caves and mines.

Legal Status: Nongame.

Long-legged Myotis
Myotis volans

Field ID: This large bat is reddish to dark brown with pale undersides; long, soft fur; and short, rounded ears. The undersides of the wings are furry from the body to about the elbow.

Size: Total length 3¾–4¼ inches (95–108 mm), wingspan 10–12 inches (25–30 cm), weight ¼–⅓ oz (7–10 g).

Habitat: Ponderosa pine forests, piñon-juniper woodlands, shrublands, canyons up to about 12,000 feet.

Distribution: From Front Range foothills west over the western two-thirds of the state.

Field Notes: The long-legged myotis is hard to distinguish from the little brown bat, with which it often shares habitat. It is slightly larger, with more rounded ears and fur on the underside of part of the wings. Emerging early in the evening, the long-legged myotis hunts over water or forest clearings no more than 15 feet above the ground. Where it inhabits deep, shadowed canyons, it may be aloft before the sun sets. It flies quickly and directly, unlike the butterfly-like flight of the canyon bat (western pipistrelle), and may stay in determined pursuit of an insect for an extended flight. It eats beetles, flies, moths, grasshoppers, dragonflies, and other insects. The long-legged myotis roosts alone or in small groups and probably hibernates in Colorado in mines and caves. It breeds and bears a single young in summer. Mothers form large nursery colonies.

Legal Status: Nongame.

White-nose Syndrome, Another Deadly Threat

A mysterious fungus is devastating entire wintering populations of some bats in the eastern United States. White-nose syndrome (WNS) is named for the white fungus that grows on the noses and other body parts of infected bats. This deadly fungus lives in cold caves where bats hibernate and infects bats while they hibernate. Though not yet fully understood, the affliction causes bats to rouse from hibernation, prematurely burn up their fat reserves, and then starve to death in mid-winter. The loss of hundreds of thousands of such beneficial animals is potentially devastating ecologically.

Though as of early 2011 WNS had not been confirmed in the western United States, it is suspected in Oklahoma, and scientists think it is only a matter of time before it infects western bat populations. Preventing transmission of the fungus from infected caves to clean ones is crucial. Recreational cavers should decontaminate clothing, boots, and equipment between cave sites. More information is available on the Colorado Parks and Wildlife website, http://wildlife.state.co.us.

Bat Conservation International maintains updated information on WNS on its website, www.batcon.org.

Yuma Myotis
Myotis yumanensis

Field ID: This medium-sized bat is pale tan to yellowish buff with darker brown wings and ears. The ears are small and the tail membrane is covered with fur.

Size: Total length 3–3½ inches (76–89 mm), wingspan 9–10 inches (23–25 cm), weight $\frac{1}{10}$–$\frac{1}{5}$ oz (3–6 g).

Habitat: Shrublands, piñon-juniper woodlands, riparian areas.

Distribution: Northwestern and far southwestern Colorado, San Luis Valley, southeastern foothills and mesas north to Colorado Springs, up to about 7,900 feet.

Field Notes: Though an animal of dry habitats, the Yuma myotis is closely tied to water, perhaps more than any other Colorado species. It often hunts along a watercourse, flying low over the water surface, or among the trees and shrubs lining a stream, pond, or lake. It feeds on aquatic insects as well as moths, grasshoppers, beetles, and flies. The Yuma myotis leaves its roost in a cave, rock crevice, or old building early in the evening, even before sunset where deep canyons lie in shadows. It may also hunt in early morning. In some parts of the country, thousands of females (who each bear a single young a year) gather into nursery colonies in caves, hollow trees, mines, and other spaces. Smaller colonies have been found in southeastern Colorado, as well as one in the Colorado National Monument. The Yuma myotis can be confused with the little brown bat, which is about the same size but darker in color.

Legal Status: Nongame.

Eastern Red Bat
Lasiurus borealis

Field ID: The fur of this medium-sized bat ranges in color from yellow to bright orange-red to brick red, with paler undersides and white-tipped fur. It has small, rounded ears, pointed wings, and a long tail. The tail membrane is furred on top.

Size: Total length 4⅛–5 inches (105–127 mm), wingspan 11–13 inches (28–33 cm), weight ¼–½ oz (7–14 g).

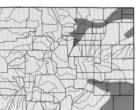

Habitat: Riparian woodlands, deciduous woodlands of farm and ranch hedgerows and windbreaks.

Distribution: Along the South Platte River from Denver to the northeast corner of the state, along the Arkansas River from Pueblo to the Kansas state line, and in the southeast corner of the state.

Field Notes: This very colorful bat has a lightly frosted look from its white-tipped fur (not to be confused with the heavily frosted hoary bat). The red bat emerges soon after sunset, flying often along the same path every night in a series of sweeping arcs through its feeding territory. It flies quickly and directly, gobbling up moths and other flying insects at the edges of woodlands, in clearings, and around streetlights. It will also land to catch beetles, ants, and crickets on the ground. During the day, it roosts, well hidden, in cottonwoods and deciduous trees, where its coloration helps it blend into the foliage. While most bats bear only one baby, the red bat has two to four young. The animals mate in the fall, but the female stores the sperm until spring, at which time eggs are fertilized and development of the young begins. Red bats are apparently rare in Colorado but because of their preference for deciduous trees, may be more numerous in the future due to the increase in riparian woodlands on the Eastern Plains and the spread of urban/ suburban forests.

Legal Status: Nongame.

Hoary Bat
Lasiurus cinereus

Field ID: This large bat has long, brown, white-tipped fur that makes it look as if it's covered with frost. It has a pale chest and belly and short, round ears. The tail membrane is heavily furred on top.

Size: Total length 4²/₃–5²/₃ inches (119–144 mm), wingspan 13–16 inches (33–41 cm), weight 1 oz (25–30 g).

Habitat: Deciduous and coniferous woodlands.

Distribution: Statewide up to about 10,000 feet.

Field Notes: Hoary means frosted, a good description of the white-tipped brown fur of this bat. Hoary bats are migratory, arriving in Colorado in April and departing in the fall. They are tree-roosting animals and may take up residence in any wooded area that provides leaf or needle cover from above but also a clear flight path to and from foraging areas. Though widespread in North America and across Colorado (and the only land mammal native to Hawaii), the hoary bat isn't abundant in our state. It hunts later in the evening than many other bats and can often be distinguished by its large size; long, slender wings; swift, direct flight pattern; and high flight. Its echolocation calls can be heard by humans as a chatter. Hoary bats prey on moths, beetles, and other large invertebrate prey, as well as small bats such as the small-footed myotis. Females usually bear two young, which are ready to fly a month after birth. Nursing young are often carried along by the female as she forages.

Legal Status: Nongame.

Silver-Haired Bat
Lasionycteris noctivagans

Field ID: This medium-sized bat has black fur tipped with silver, white or yellow. Its ears are fairly short, naked, and rounded. The tail membrane is partly furred on the top.

Size: Total length 3½–4⅓ inches (89–110 mm), wingspan 11–13 inches (28–33 cm), weight ⅓–⅖ oz (9–11 g).

Habitat: Scattered tree stands and forest edges, caves, buildings, and wood piles.

Distribution: Statewide up to about 9,500 feet.

Field Notes: The silvery fur of this bat makes it quite a striking, handsome animal. The white-tipped hairs give it a frosted appearance. Its scientific name means "shaggy, night-wandering bat." A bat of woodlands, it roosts in tree cavities and the little crevices beneath loose tree bark, though it also inhabits open caves or mines and the occasional building. It seems to avoid enclosed spaces such as attics. Silver-haired bats forage along streams and around ponds lined with trees, emerging fairly late, after other bats are already abroad. Its slow flight, unusual for bats, is fairly straight and close to the water surface or just a few feet above the ground. Occasionally one is snagged by the line of a fly fisher. The silver-haired bat is a solitary creature and is common throughout temperate North America. A few silver-haired bats have been found hibernating in mines, but it is likely most of them migrate south during the cold months. Mated females store sperm over winter and bear two young between June and mid-July.

Legal Status: Nongame.

Identifying Colorado Bats In Flight

Most of us will never see a bat except as a silent shape fluttering in the moonlight. Knowing the flight patterns and foraging habits of each species can help identify them.

Species	Time of flight
California Myotis	Early evening. Leaves roost right after sunset.
Western Small-footed Myotis	Leaves roost very early in evening.
Long-eared Myotis	Late evening, though earlier at higher elevations.
Little Brown Bat	Emerges at dusk.
Fringed Myotis	Emerges about two hours after sunset.
Long-legged Myotis	Emerges in early evening, before dark.
Yuma Myotis	Emerges very early in evening or late afternoon in shadowed canyons.
Eastern Red Bat	Emerges in early evening.
Hoary Bat	Emerges late evening.
Silver-haired Bat	Emerges late, after other bats.
Canyon Bat (Western Pipistrelle)	Emerges very early evening, sometimes before sundown, and again in early morning.
Big Brown Bat	Emerges at dusk.
Spotted Bat	Emerges late, well after dark and active still at midnight.
Townsend's Big-eared Bat	Emerges well after dark.
Pallid Bat	Emerges late.
Brazilian Free-tailed Bat	Emerges at dusk. Outflights can include hundreds or thousands of bats.
Big Free-tailed Bat	Emerges late evening when nearly dark.

Flight/Foraging pattern

Forages low—6 to 9 feet above the ground, near trees.

Slow, fluttering, highly maneuverable flight. Feeds low among trees or over brush.

Hovers at foliage along edge of trees or in forest clearings.

Zigzag flight along a repeated circuit. Forages over water or ground or among trees, 9 to 18 feet above the ground.

Slow flight. Forages along water, above shrubs and woodlands, or low over meadows. Hovers at foliage.

Quick, direct-pursuit flight. Forages over ponds, streams, open meadows, or forest clearings, 9 to 12 feet above the ground, on a repetitive circuit.

Forages near water, often along a watercourse.

Steady, rapid flight repeated along same route. High, lazy flight pattern in sweeping arcs above the trees in early evening; close to or on the ground after dark. Often forages around lights.

Fast, direct flight with long, slow wingbeats.

Very slow, straight flight on a repeated circuit. Forages in and near woodlands near water, close above the ground or water.

Slow, erratic, fluttery, butterfly-like flight. Forages along canyon walls and over scattered boulders and shrubs.

Strong, straight, direct flight 20 to 30 feet above the ground. Forages in the open.

Forages in the open, 15 to 30 feet above the ground.

Swift, highly maneuverable flight. Forages along the edge of vegetation.

Flies low, foraging 6 to 36 inches above the ground. Drops abruptly to the ground to capture prey.

High, fast, forceful flight on long, angular, narrow wings. Flies long distances to feeding grounds.

Strong, powerful flight.

Canyon Bat/Western Pipistrelle
Pipistrellus hesperus

Field ID: This tiny bat is pale gray or yellowish, with short, black ears and a distinctive dark mask on the face.

Size: Total length 2⅓–3⅓ inches (60–85 mm), wingspan 7–9 inches (18–23 cm), weight ¹⁄₁₀–⅕ oz (3–6 g).

Habitat: Canyons, semidesert shrublands, mesas, deserts.

Distribution: Far western Colorado and southeastern Colorado.

Field Notes: A tiny bat fluttering butterfly-like in the early evening air is likely a canyon bat, previously known as a western pipistrelle. This handsome bat, the smallest in Colorado, is distinguished by tiny size, pale color, and slow, erratic flight. A bat of arid landscapes, it is found in canyons and desert shrublands as well as atop dry mesas, always near water. It doesn't seem to use human-made structures, caves or mines as roosts, preferring hidden spots in thick foliage, beneath rocks, in crevices or even animal burrows. It emerges very early in the evening, often while there is still daylight, fluttering around the shadowed airspaces of canyons from 6 to 45 feet above the ground. Flying into a swarm of insects, it can quickly consume 20 percent of its body weight. The canyon bat limits its hunting activity to still evenings, probably because it's too small to fight much of a breeze. It doesn't migrate but hibernates in caves and mines near its summer range. Females bear one to two young in June or July and may form small nursery colonies or remain solitary. There is a single record of the eastern pipistrelle (now known as the tri-colored bat) in Colorado.

Legal Status: Nongame.

Big Brown Bat
Eptesicus fuscus

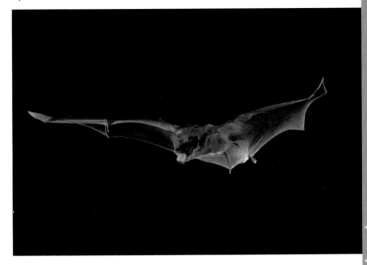

Field ID: This large bat ranges from pale to reddish to very dark brown. It has dark ears, wings, and tail; medium-length, rounded ears; and long, pointed wings. The tragus is rounded and only half the length of the ear.

Size: Total length 3½–5⅓ inches (89–135 mm), wingspan 13–16 inches (33–41 cm), weight ⅖–⅗ oz (11–17 g).

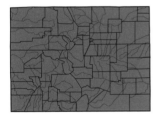

Habitat: Forests, shrublands, riparian areas, urban/suburban neighborhoods, agricultural lands.

Distribution: Statewide.

Field Notes: This large, common bat is distributed throughout North America. It adapts well to humans, roosting in urban and suburban buildings, houses, attics, under bridges as well as hollow trees, caves, rock crevices, or any sheltered spot. Because of its size, distribution, and tolerance of humans, it is the bat most frequently seen, foraging in parks, backyards, and around lampposts. Big brown bats leave day roosts at dusk, hunting in a straight, determined flight 20 to 30 feet above the ground. Groups of them hunt over the same spots week after week, feeding on beetles, wasps, flies, and flying ants. Mated females store sperm through winter, bearing a single young in summer. They gather in nursery colonies of up to hundreds of mothers and nursing young, in caves or attics. If a baby falls to the ground, the mother picks it up and carries it back to the roost site. If mothers are disturbed, they relocate their babies to a quieter spot. Young are able to fly in a month. Big brown bats remain in Colorado in winter, hibernating in caves, mines, and rock crevices.

Legal Status: Nongame.

Shh! Bats Sleeping

Bats are highly beneficial animals, feeding on untold millions of mosquitoes and other insects each year. But these small, fragile creatures are easily injured. They are also slow-reproducing. The females of most species produce only one baby a season. So, if you discover a bat roost, don't harass or disturb the animals. The website for Bat Conservation International, www .batcon.org, has information on safely relocating bats and excluding them from buildings.

Bats are especially vulnerable in winter, when they hibernate to survive the cold months, relying on stored body fat. If a bat is disturbed and rouses from torpor, it can easily burn up so much of its energy reserve that it dies before spring.

If you encounter hibernating bats in a cave, abandoned building, or anywhere, back off and don't disturb them. It really is a matter of life or death.

Spotted Bat
Euderma maculatum

Field ID: Perhaps the rarest of Colorado's bats, this is also the most distinctive in appearance. It is fairly large, with a nearly one-foot wingspan; huge, pink ears; and a white-frosted black coat with three large white spots on the back—one on the rump and one at the base of each ear.

Size: Total length 4¹/₈–4²/₃ inches (105–119 mm), wingspan 13–15 inches (33–38 cm), weight ½ oz (13–14 g).

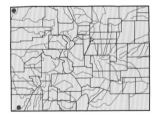

Habitat: Dry shrublands, ponderosa pine and piñon-juniper woodlands, rocky cliffs near water.

Distribution: Far northwestern Colorado in and near Dinosaur National Monument and in Mesa Verde National Park in southwestern Colorado.

Field Notes: The spotted bat is not only rare but also one of the most distinctive and beautiful bats in Colorado. The large, white spots on its black back look somewhat like a face. This bat is known only from a handful of documented records from Browns Park, and from several unofficial sightings. A denizen mainly of the Southwest, it likely inhabits canyons of western Colorado at lower elevations. Because it is difficult to capture, the spotted bat may be more common than previously thought. The spotted bat's echolocation calls are low in pitch and can be heard as clicks by humans. It eats beetles and grasshoppers, but its preferred food is moths. However, it apparently likes only the abdomens and will tear off the head and wings. The spotted bat crawls easily across flat surfaces. Females bear a single young in summer. Any sightings of spotted bats should be reported to Colorado Parks and Wildlife.

Legal Status: Nongame.

Townsend's Big-eared Bat
Corynorhinus townsendii

Field ID: This pale brown to dark gray bat has extremely long ears with ridges across them, a glandular lump on either side of the muzzle, and an unfurred tail membrane. The slender tragus is rounded and about half the length of the ear.

Size: Total length 3½–4⅓ inches (89–110 mm), wingspan 12–13 inches (30–33 cm), weight ⅓–½ oz (9–14 g).

Habitat: Shrublands, piñon-juniper woodlands, open coniferous forests.

Distribution: From eastern foothills west across the state up to about 9,500 feet and in far south-southeastern Colorado.

Field Notes: This bat's exceptionally long ears dwarf its small face. The ears are flexible. Laid back, they reach the middle of the body. These bats emerge well after dark, feeding on moths, flies, and caddis flies, and gleaning insects from vegetation. They listen for the footsteps and chewing sounds of caterpillars and insects, then move in to pluck them off of vegetation. Mated females store sperm until they emerge from hibernation, when fertilization occurs. A single young is born in May or June. Females gather by the hundreds in nursery colonies. When they leave their young to forage, the huddled mass of bodies warms and protects the young. Instead of squeezing into tiny cracks, Townsend's bats roost in large, open areas such as mines and caves. This leaves them vulnerable to disturbance, and their numbers have declined in many areas. If disturbed during hibernation, they may burn crucial stored fat and not survive winter. Even without disturbance, they lose half their body weight through winter and many do not emerge in spring.

Legal Status: Species of special concern.

Pallid Bat
Antrozous pallidus

Field ID: This large, pale bat ranges from cream to light brown. A series of warty, glandular lumps form a ridge around the nostrils that make the muzzle blunt. The big, broad ears can be an inch long and are crossed by a series of lines.

Size: Total length 3½–4⅖ inches (89–112 mm), wingspan 15–16 inches (38–41 cm), weight ¾–1¼ oz (21–34 g).

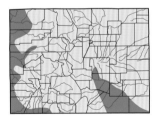

Habitat: Rocky areas of shrublands, grasslands, deserts and piñon-juniper woodlands up to about 7,000 feet.

Distribution: Far western Colorado and along the Colorado and Gunnison rivers, canyons and mesas of the southeastern part of the state, north along the eastern foothills to Colorado Springs.

Field Notes: Pallid bats head out to hunt about an hour after dark. They capture flying insects in the air but also, in most unbat-like fashion, land to catch prey on the ground. They consume many beetles but also take scorpions, crickets, and even small lizards, mice, and other bats. They apparently locate terrestrial prey by sight, then land and crawl after it, moving on the outer part of the wrists, with their wings folded behind. Hunting on the ground may increase their prey base but also their chance of being eaten by something else. Once they seize an insect, from the air or ground, they fly to a feeding roost, where they hang upside down and tear off the wings and other hard, inedible parts, which accumulate on the ground below the roost. By day, they roost in rock crevices. Pallid bats probably hibernate in Colorado. Females hang upright to give birth and catch their two babies in the tail membrane. Young can fly in six weeks.

Legal Status: Nongame.

Family Molossidae—Free-tailed Bats

The tails of molossid bats extend well beyond the interfemoral, or tail, membrane; hence the name "free-tailed."

These bats have short, dark fur and a musty odor. They tend to live in colonies, usually in caves but also in abandoned structures. The bats that make the spectacular evening outflights of thousands of individuals at Carlsbad Caverns National Park and from the Orient Mine in the San Luis Valley are Brazilian free-tailed bats.

Animal Sign: Flight pattern (see Identifying Colorado Bats in Flight sidebar on pages 48–49); guano (droppings) accumulated in roost caves, but not usually identifiable by species.

Brazilian Free-tailed Bat
Tadarida brasiliensis

Field ID: This large, gray-brown bat is paler underneath, with black wings and short, black ears that almost join at the middle of the forehead. The ears and wings are thick and leathery. Most of the tail is not connected to the tail membrane.

Size: Total length 3½–4 inches (89–102 mm), wingspan 12–14 inches (30–35 cm), weight ¼–⅖ oz (8–12 g).

Habitat: Shrublands, grasslands, piñon-juniper woodlands.

Distribution: From central-western and southwestern Colorado eastward through the San Luis Valley and into far south-southeastern Colorado.

Field Notes: This bat is the legendary species of Carlsbad Caverns National Park, though its numbers are declining. Its enormous colonies stage a spectacular fly-out at dusk as the adults leave to hunt. A colony estimated at a quarter million individuals inhabits the Orient Mine in the San Luis Valley. These bats have long, narrow wings and fly quickly and aggressively, foraging over a broad area up to 40 miles from their roost. Males and females overwinter together. They breed in late winter and early spring, and eggs are fertilized at breeding. Females gather in large nursery colonies in mines and caves, under bridges, in buildings or any dark, dry roost with a clear space below that the bats can drop down into as they take flight. The mother leaves her single baby at a spot on the cave ceiling and goes out hunting. Amazingly, she can locate her own baby within the teeming mass of thousands or millions of bats. Within about a month, the babies are covered with fur, adult-size, and ready to hunt on their own. These bats eat moths, mosquitoes, and other insects, a large colony consuming tons of insects each year. They migrate south for winter.

Legal Status: Nongame.

Big Free-tailed Bat
Nyctinomops macrotis

Field ID: This large bat is pale brown to reddish to black, with paler undersides. The large, rounded ears join at the middle of the forehead. Most of the tail is not connected to the tail membrane.

Size: Total length 4⅘–5½ in (122–140mm), wingspan 17–18 in (43–46 cm), weight 1 oz (25–30 g).

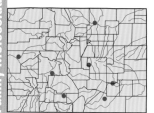

Habitat: Canyons, cliff faces.

Distribution: Through central and western Colorado (based on a handful of records).

Field Notes: Once thought to be just an occasional wanderer into the state, new evidence suggests this species breeds in western Colorado. The big free-tailed bat is well named, being considerably larger than the Brazilian, with a wingspan of nearly a foot-and-a-half. These bats are strong, powerful fliers, emerging late in the evening to hunt high in the sky, the sound of their large, leathery wings discernible. Their echolocation calls are also audible to humans. They inhabit crevices and cracks along cliffs and canyons. During daytime, colonies within horizontal rock crevices can often be heard chattering, the sound, described as similar to the cooing of doves, echoing off the rock walls. They eat moths and flying insects and probably pick ground-dwelling prey, such as crickets, grasshoppers, and ants, off the sides of cliffs. Females gather in small nursery colonies and bear a single young in June or July.

Legal Status: Nongame.

ORDER
CINGULATA

Armadillos: Armored Mammals

Armadillos were formerly grouped with aardvarks and sloths into the order Edentata but are now classified within their own order: Cingulata. This name derives from the Medieval Latin word *cingulum*, meaning "a girdle, belt or sash," a reference to the armor bands encircling the armadillo's body.

Dasypodidae is the only surviving family of this order (there are five extinct families), with 20 living species. As with opossums, all the species of this order inhabit South America except for the nine-banded armadillo.

The characteristics of this order are a carapace of bony plates making a protective covering for the body, scant hair, and teeth that are rootless, ever-growing, and lack enamel. Armadillos live primarily on the ground, or in the ground, and are well adapted for digging.

Family Dasypodidae—Armadillos

Some twenty species of armadillos are distributed from South America through Central America into Mexico, but only the nine-banded armadillo inhabits the United States. The species' range has been expanding from south Texas northward over the last century, up into the central Great Plains as far as Nebraska and east to South Carolina and Florida. Armadillos have been reported in southern Illinois and Indiana.

Armadillos vary in size from the five- to six-inch pink fairy armadillo weighing only four ounces, to the giant armadillo, a behemoth that can measure as much as five feet long and weigh up to 70 pounds.

An armor made of horny plates covers the back, sides, tail, and parts of the head of armadillos. Some species also have hair growing between the plates. The front feet have two or three enlarged toes with long, curved claws. Armadillos are strong diggers, using their powerful front legs and claws to scratch and dig after insects, carrion, small reptiles, mammals, and, for some species, plant matter. Their teeth are very reduced, with no incisors or canines and undifferentiated cheek teeth that are like small pegs.

Animal Sign: Disturbed soil and rootings in search of insects; excavated anthills; trundling track in loose dirt.

Nine-banded Armadillo
Dasypus novemcinctus

Field ID: This housecat-sized animal has a rounded back covered with leathery, bony armor plates, a pointed snout, naked ears, and a short, tapered, scaly tail. It is a mottled dark to grayish ivory.

Size: Head/body length 23½–32 inches (60–81 cm), tail 9⅓–14½ inches (24–37 cm), weight 6½–15 lbs (3–7 kg).

Habitat: Deciduous woodlands along rivers and streams.

Distribution: Eastern Colorado in counties bordering Kansas and Oklahoma.

Field Notes: "Tank" is a good nickname for these armor-plated beasts that trundle along in search of food. The word *armadillo* is Spanish for "little armored one." Armadillos are outfitted with a horny armor or shell, called a carapace (the same term used for a turtle's shell), on their backs and sides. The carapace is made of bony plates (scutes) covered with a horny skin. Bands of flexible skin between the plates allow the animal to move more freely. When threatened, an armadillo adopts the bowling ball defense, rolling into a tight, round shape with head, feet, and tail tucked, protecting its soft belly and leaving only its armored back exposed. The legs, top of head, and tail are also armored. Armadillos can run fairly fast over short distances. Fleeing into their burrows, they brace their legs against the walls with their armored backs facing the hole and are almost impossible to remove.

Armadillos don't hibernate; instead, they shelter from cold weather in deep burrows. They breed in late summer, but embryo implantation is delayed until early winter. A litter of usually four young is born after about four months. The babies are identical—quadruplets all developed from a single fertilized egg. The fist-sized newborns are born with their eyes open and their armored carapace formed but soft. They are able to walk within a day, though they will not be weaned from dependence on their mother for food until they're about three months old. Colorado boasts only a handful of records of armadillos, all from the eastern part of the state and with no confirmed breeding, but it is likely this adaptable animal will continue to expand its range, with increased sightings in Colorado.

Legal Status: Nongame.

ORDER
LAGOMORPHA

Rabbits and Relatives: Pikas, Rabbits, and Hares

Rabbits, hares, and pikas are often mistakenly identified as rodents, but they belong to their own family. Lagomorphs (which means "rabbit-shaped") have an interesting physical trait—a second pair of upper incisors set just behind the primary teeth. Like rodents, they lack canine teeth and have a space, a diastema, between incisors and cheek teeth.

Unlike rodents, lagomorphs can't grasp food with their front paws. They have five toes on the front foot and four on the hind. Lagomorphs have fully furred paws, short tails (the pika's is not externally visible), and large (to huge) ears. Even the pika's rounded ears are large for its body mass.

To complete digestion of the plant material they eat, lagomorphs practice coprophagy. They produce two types of fecal pellets—a soft pellet that is reingested to extract more nutrients, and a hard waste pellet. Reprocessing fecal pellets can yield five times as many nutrients as on the first pass through the digestive tract.

Lagomorphs do not hibernate but remain active through winter. Several species molt from mottled brown summer pelage to a winter coat of pure white.

Family Ochotonidae—Pikas

Pikas have evolved a highly specialized lifestyle, living year-round at or above timberline. There are two species of this family distributed through high mountain regions of western North America. The collared pika is restricted to Canada and Alaska.

Pikas have rounded bodies with no noticeable tail, and large, rounded ears. They don't hibernate, relying on thick fur and the insulating cover of heavy snow to survive winter in the harsh alpine environment. Pikas are among the first animals being impacted by the effects of global climate change as the high altitude snowpacks they depend on for winter protection are being reduced and are melting earlier. Current (2012) studies indicate that Colorado pika populations are still strong, but time will tell whether these alpine-dwellers will suffer from the effects of climate change even in our state, which is a stronghold for the species.

Animal Sign: High-pitched, squeaky barks; rounded tufts of dried grass and vegetation stuffed between the rocks of a talus slope; very small, rounded black pellets scattered among the rocks.

American Pika
Ochotona princeps

Field ID: These cousins of rabbits are small and grayish brown, with short, round ears and no visible tail.

Size: Head/body length 6⅓–8½ inches (16–22 cm), weight 4¼–6⅓ oz (113–180 g).

Habitat: Rocky talus slopes at and above timberline, open areas amid sub-alpine and timberline forests.

Distribution: Throughout the higher mountains of the state, above 10,000 feet.

Field Notes: Pikas live at the roof of Colorado, at and above timberline. Despite the harshness of alpine life, they don't hibernate but remain active, sheltering in rock dens insulated beneath the cover of snow. They spend much time in the summer cutting wildflowers and grasses, which they store as winter food in little haystacks between the rocks. They even turn their "hay" so it cures properly, leading to one of the pika's nicknames, "rock farmer." Though pikas have short, rounded ears and look like rodents, they are actually cousins of rabbits. Pikas are extremely well-camouflaged and difficult to spot against the rocky talus, but they are highly territorial and will challenge hikers with repetitive, high-pitched barks— chirk! chirk! chirk! Pikas breed in spring, producing one litter of two to five young, born in 30 days. Young are mature after the next spring. Pika populations are threatened by global climate change and atmospheric warming. These changes encourage the upslope migration of lower-elevation plants and animals into pika habitat, reduce critical insulating snow cover in winter, and lead to potentially lethal elevations of body temperature. Local extinctions of pika populations have already been reported in some areas outside Colorado.

Legal Status: Nongame.

Family Leporidae—Rabbits and Hares

Rabbits and hares are among the most familiar and identifiable mammals. They occupy a variety of habitats statewide and are generally instantly recognizable by their very long legs, feet, and ears and short, fluffy tails. Their eyes are large and located on the side of the head.

Behaviorally, rabbits and hares freeze if they detect danger or a disturbance, then react explosively, running very fast into cover. Extremely long legs and feet are an adaptation for running. As the animal pushes off, its long legs and feet give it great extension, providing more distance with each stride and for the energy invested.

Animal Sign: Round, somewhat flattened brown pellets scattered on the ground; tracks—four toes show in both front and hind tracks (fifth front toe rarely shows); elongated hind track; hind feet show in front of forefeet.

Desert Cottontail
Sylvilagus audubonii

Field ID: This small rabbit has a rough, brown-gray coat flecked with black, with paler sides, white underparts, and a buff-brown area on the throat and chest. Its large ears are scantily haired. It has large hind feet and a fluffy tail like a puff of cotton.

Size: Head/body length 12–15 inches (30–38 cm), ear 3–4 inches (76–102 mm), hind foot 2¾–3½ inches (70–89 mm), weight 1½–2⅔ lbs (600–1,200 g).

Habitat: Grasslands, shrublands, open mountain forests.

Distribution: Eastern, southern, and far western Colorado, up to about 7,000 feet.

Field Notes: This is a rabbit of short-grass prairie and dry shrublands. It is most active early and late in the day, spending the hottest hours resting beneath a shrub or in the burrow of another animal. In some areas, the desert cottontail is nicknamed "prairie dog rabbit" because it is so common in the colonies of these ground squirrels. Relying upon speed to escape predators, the desert cottontail can run up to 15 miles per hour. It can also swim and will climb small shrubs or the sloping trunks of trees to look out for danger. It feeds on grasses, wildflowers, and forbs and is active throughout the year. Females dig shallow nest chambers for their litters of up to six young. The nests are lined with grass and filled with soft fur. The babies mature quickly and are ready to mate by about three months old. Females usually have two to four litters a year.

Legal Status: Small game.

Eastern Cottontail
Sylvilagus floridanus

Field ID: This large cottontail is dark gray washed with red and flecked with black. It has relatively long ears, large, whitish hind feet and a fluffy, cottonball-like tail.

Size: Head/body length 14–17 inches (36–43 cm), ear 2½–3 inches (64–76 mm), hind foot 3–4⅓ inches (76–110 mm), weight 2–4 lbs (900–1,800 g).

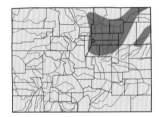

Habitat: Grasslands, pastures, edges of farm fields, open shrublands, riparian areas, suburban parks, gardens and yards.

Distribution: Northern Colorado, especially the South Platte River valley.

Field Notes: The eastern cottontail is larger and darker than the desert or mountain cottontails, with shorter ears relative to its body. It feeds on a variety of grasses, flowers, and forbs (as many gardeners can attest), switching to the buds, bark, twigs, and shoots of shrubs and trees in winter. It is nocturnal and crepuscular. As the most common rabbit in North America, it adapts well to life around people. Humans think rabbits are docile, but they have their own elaborate society with dominance, aggressive interactions, and courtship between males and females. Except for during the breeding season and when a female has young, they are solitary. Eastern cottontails breed from late winter through August. Like most rabbits, they are prolific breeders, with three to eight litters a year averaging three or four young. Cottontails are an important prey animal for avian and mammalian predators.

Legal Status: Small game.

Mountain Cottontail/Nuttall's Cottontail
Sylvilagus nuttallii

Field ID: This small rabbit is dark reddish brown, with black-tipped hairs. The underparts and "cotton" tail are white. The ears are furry and shorter than those of other cottontails.

Size: Head/body length 12–14 inches (30–36 cm), ear 2 $\frac{1}{5}$–2$\frac{3}{5}$ inches (56–66 mm), hind foot 3½–4$\frac{1}{3}$ inches (89–110 mm), weight 1½–3 lbs (700–1,400 g).

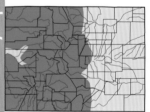

Habitat: Mountain shrublands, mountain parks, open forests of foothills and mountains.

Distribution: Central and western Colorado from the foothills of the Front Range to Utah, between 6,000 and 11,500 feet.

Field Notes: Colorado's three cottontails are similar in appearance but occupy different ranges and habitats. The mountain cottontail, also called Nuttall's cottontail after a famous naturalist, is true to its name, inhabiting mountain parks and shrublands and the edges of montane forests and piñon-juniper woodlands. In summer, it feeds on grasses, wildflowers, and forbs, switching to big sagebrush, rabbitbrush, juniper, and other shrubs for winter. It is darker and has shorter ears and hind legs than the desert cottontail, with whom it shares areas of western Colorado. Over most of its range it is the only cottontail. Like its cousins, it is nocturnal, sheltering during the day in scrapes beneath shrubs, fallen logs, old burrows, or other cover. Mountain cottontails breed from February through July, producing several litters of four to five young a year. It builds nests beneath logs or other shelter, lining them with soft plant fibers and rabbit fur. As late summer drought dries the vegetation, mountain cottontails cease their reproductive season.

Legal Status: Small game.

Rabbit or Hare?

People use the terms *rabbit* and *hare* almost interchangeably, but there is a difference. The animals we know as jackrabbits are actually hares, while cottontails are rabbits. Domestic rabbits, though sometimes called Belgian hares, are true rabbits.

Rabbits

Born without fur, helpless, and unable to see
Dig burrows for shelter
Are smaller and not as fast
Have shorter ears and feet
Molt into a grayish winter coat

Hares

Born more mature, with fur and open eyes
Take cover in depressions beneath shrubs but rarely excavate a den
Are generally larger and faster
Have longer ears, longer legs, and larger feet
Molt into a white winter coat (not all species)

Snowshoe Hare
Lepus americanus

Field ID: This medium-sized hare has very large hind feet and moderately long ears. It is rusty to gray-brown in summer, with white underparts, legs, and feet. In winter it molts to pure white.

Size: Head/body length 13–18 inches (33–46 cm), ear 3½–4 inches (89–102 mm), hind foot 4⅓–6 inches (110–152 mm), weight 2–4 lbs (900–1,800 g).

Habitat: Montane and subalpine forests, mountain willow thickets, alpine tundra near tree line.

Distribution: Throughout the mountainous region of central Colorado, from about 8,000 to 11,500 feet.

Field Notes: Snowshoe hares are wonderfully adapted to their high mountain habitat, particularly for winter. Their outsized hind feet allow them to move around in deep snow by acting as "snowshoes" to displace their weight. They transform from a summer coat of rusty brown to pure white, which magically conceals them (except for their black eyes) against winter's snowy backdrop. They feed on the leaves, needles, buds, twigs, and bark of conifers and deciduous trees and mountain shrubs, as well as grasses, wildflowers, and forbs. Their feeding activity during times of deep snow can be seen as a browse line as much as six feet high on trees and shrubs. Snowshoe hares are nocturnal, secreting themselves during the day in shallow scrapes set within dense thickets of conifers or shrubs. They breed from April to September. Females bear two to three litters of one to seven young. The onset and length of breeding and litter size are affected by the amount of winter food available. The presence of healthy snowshoe hare populations is an important factor in the restoration of lynx to the Colorado mountains.

Legal Status: Small game.

Black-tailed Jackrabbit
Lepus californicus

Field ID: Jackrabbits are grayish black with white underparts. The tail has a black stripe on the top. They have long legs, long ears, and bulging eyes.

Size: Head/body length 17–21 inches (43–53 cm), ear 6–7 inches (15–18 cm), hind foot $4^{1}/_{3}$–$5^{2}/_{3}$ inches (110–145 mm), weight 3–7 lbs (1.4–3.1 kg).

Habitat: Grasslands, pastures, meadows and semidesert shrublands below 7,000 feet.

Distribution: Eastern Colorado from the Eastern Slope of the Rockies to the Kansas state line, and far western Colorado.

Field Notes: Jackrabbits are legendary for their explosive, zigzag running style. They have longer legs and bigger feet than cottontails, which give them a longer stride and thus greater speed. As the food of choice for many predators, jackrabbits need to be fast, and prolific. They breed from February to July. One female may produce up to 30 young a year from as many as five litters. A fourth of female young are born early enough to breed themselves the same season. Though called a jackrabbit, this animal is actually a hare. It is mainly nocturnal and crepuscular, feeding on grasses, sedges and wildflowers, switching to bark, twigs, and buds of trees and shrubs in winter. Jackrabbits may gather in winter in groups of 30 or more. Studies have found that the populations of coyotes are closely tied to those of jackrabbits. When jackrabbit populations decline, coyote predation on livestock goes up.

Legal Status: Small game.

White-tailed Jackrabbit
Lepus townsendii

Field ID: In summer, this large hare is soft, pale brownish gray, with a white rump and tail and long hind legs. Its large ears are white with black tips. In winter, it turns all white, with black ear tips.

Size: Head/body length 18–22 inches (46–56 cm), ear 5–6 inches (13–15 cm), hind foot $5^2/_3$–$6^4/_5$ inches (145–173 mm), weight 5–10 lbs (2.2–4.5 kg).

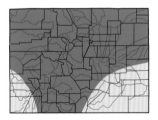

Habitat: Grasslands, mountain parks, alpine tundra, semi-desert shrublands.

Distribution: Statewide except for extreme southwestern and southeastern Colorado.

Field Notes: This is an animal of open grasslands and mountain parks, where its ability to dart and change direction on a dime is a vital skill. Like its close relative, the black-tailed jackrabbit, it is a sprinter, eluding predators with bursts of speed. The white-tailed jackrabbit can reach 40 miles per hour and leap a horizontal distance of 17 feet. It is crepuscular and nocturnal, sheltering during the day in shallow scrapes beneath shrubs, where it relies upon its camouflaging coat to escape detection. It may sit tight until approached very closely, then burst from cover. This "jackrabbit start" not only provides a head start but also startles and confuses predators. White-tailed jackrabbits will tunnel through deep snow to forage. They feed on grasses and sedges in summer, switching to forbs and shrubs in winter. While most hares don't use burrows, white-tailed females may bear their young in nests hidden below ground. They may have multiple litters a year, but mountain populations have only one, perhaps due to reduced resources and a shorter growing season. This is by far the largest lagomorph in Colorado, with a broad distribution from the Eastern Plains to the high mountains up to 14,000 feet and into western valleys.

Legal Status: Small game.

ORDER
RODENTIA

Rodents: Gnawing Mammals

Rodents are the largest order of mammals in both Colorado and the world, comprising 40 percent of all mammal species worldwide.

Renowned for their gnawing, rodents have ever-growing top and bottom incisors. Only the front surface of the incisors has enamel (colored orange or brown). Because the back surface wears more than the enameled front, the animal's chewing keeps the teeth sharp and chisel-shaped. Rodents lack canine teeth and there is a large space, the diastema, between the incisors and cheek teeth. The rodent jaw moves front-to-back, rather than side-to-side as with most mammals. The structure of the jaw allows rodents to independently gnaw with the front teeth or chew with the molars.

Beyond their gnawing ability, rodents are a diverse group with species adapted for life underground, at the ground surface, in trees, and on water. Many rodents have dexterous forepaws for grasping food while they gnaw. Most have four toes on the front foot and five on the hind. They tend to be small animals, but the beaver, Colorado's largest rodent, can weigh up to 70 pounds. Though primarily herbivorous, most rodents occasionally eat animal flesh, mainly insects and carrion, and the grasshopper mouse is significantly carnivorous.

Family Sciuridae—Squirrels

Like off-the-rack clothing, this diverse family comes in small (chipmunks), medium (ground and tree squirrels), large (prairie dogs), and extra-large (marmots). The smallest (the diminutive least chipmunk) weighs 1 ounce, while the largest (the yellow-bellied marmot) weighs in at a plump 10 pounds. Squirrels make their homes in places as diverse as the animals themselves, from underground burrows to the tops of trees.

Because they are generally diurnal, very active, and not secretive, sciurids are easy to see and entertaining to watch. These highly communicative animals spend a great deal of time running, climbing, calling, and waving their tails in full view, relying for protection on quickness and multiple refuges rather than stealth. Many are very social and live communally, with behaviors and vocalizations that are fun to watch, or hear, and interpret. Many squirrels are also either highly patterned, like chipmunks and thirteen-lined ground squirrels, or colorful, like Abert's squirrels and marmots.

Most of this family either enters full hibernation in winter or has bouts of torpor through the cold months. They have four toes on the front foot and five on the back. Their tails range from short to very long, and thinly haired to fluffy.

Animal Sign: Small "handlike" tracks showing four toes on the front track, five on the back; hind feet placed in front of front feet; burrows (appropriate to size of squirrel); dry scat pellets; prairie dogs—mounds around burrows, often with "nose prints"; pine squirrels—middens of cone debris; fox squirrels—nest is a ball of leaves set high among branches of deciduous trees.

Cliff Chipmunk
Tamias dorsalis

Field ID: This relatively large chipmunk is grayish, with faint stripes on the back and whitish underparts. It has a long, somewhat fluffy tail.

Size: Head/body length 5–6 inches (13–15 cm), tail 3½–4⅕ inches (89–107 mm), weight 2–3 oz (57–85 g).

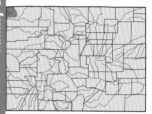

Habitat: Rocky outcrops and cliffs within ponderosa pine and piñon-juniper woodlands.

Distribution: Far northwestern corner of Moffat County.

Field Notes: Don't expect this drab chipmunk to beg at your picnic. It is shy and retiring, preferring to stay close to its den within a crevice in the rocks or under a boulder. Its camouflaged coat allows it to blend into its surroundings. The cliff chipmunk emerges from its den to forage for food from early to mid-morning, then again in late afternoon. It feeds on seeds, flowers, juniper berries, and vegetation. Good at climbing, it might be spotted in the top of a juniper tree feasting on berries, which it caches along with seeds as a future food supply. These chipmunks enter winter dormancy, rousing occasionally on warm days when they recharge their energy stores by feeding on their cached food. They mate in April and May and bear four to eight young. They may mate again in late summer.

Legal Status: Nongame.

Least Chipmunk
Neotamias minimus

Field ID: This smallest of Colorado chipmunks has the crispest contrast in body stripes. It has a brownish body and five dark stripes alternating with four pale stripes down the back. The center stripe goes from head to tail. There are alternating dark and pale stripes on the side of the face and across the eyes. The tail is long and slightly fluffy.

Size: Head/body length 3²/₃–4½ inches (93–114 mm), tail 3–4½ inches (76–114mm), weight 1–2 oz (28–57 g).

Habitat: Shrublands, piñon-juniper woodlands.

Distribution: From the eastern foothills west across the state.

Field Notes: This small, busy, brightly colored creature is the familiar chipmunk found throughout the central mountains of Colorado. It dashes around in short bursts, its tail carried straight up. It feeds on flowers, seeds, pine nuts, berries, and other plant material, as well as insects and carrion. Least chipmunks dig burrows under logs, rocks, and tree roots. They often sit atop a favorite rock near the den to feed or watch for danger. The smallest, most common, and most broadly distributed of Colorado chipmunks, the least chipmunk is found throughout the western two-thirds of the state, in a variety of habitats. Countless summer picnickers have made this active chipmunk's acquaintance in parks and picnic areas where handouts from people have acclimated it to humans. It doesn't hibernate but enters winter dormancy, rousing occasionally to feed on stored food. Least chipmunks breed March through June, producing a single litter of four to six young per year.

Legal Status: Nongame.

Mooches

It's hard to picnic in the Colorado mountains without attracting a begging chipmunk or ground squirrel. But no matter how cute the animal, resist the urge to feed it.

Wildlife find the nutrients they need in wild food. Human food contains salt, sugar and chemical additives that are unhealthy for wild animals. Feeding wildlife leads them to overcome their natural fear of humans and can lead to behavior that is unnatural and unsafe for both wildlife and people. (The exception is feeding birds in your backyard.) The time animals spend begging reduces the amount of time they would otherwise spend foraging naturally.

Chipmunks and small mammals that are after human food may bite, scratch, and destroy property. Large mammals like bighorn sheep may approach people and damage cars or other property. It's illegal under state law to feed game species. And seeing animals begging for handouts is not a very rewarding wildlife watching experience.

So, when a wildlife mooch approaches you for a handout, just say no. It's best for the animal.

Colorado Chipmunk
Neotamias quadrivittatus

Field ID: This large chipmunk is reddish brown with yellowish buffy sides and a distinctive black stripe down the center of the back. It has alternating pale and reddish brown stripes on the sides of the back and a long, fairly bushy tail.

Size: Head/body length 4½–5 inches (114–127 mm), tail 3⅕–4½ inches (81–114 mm), weight 2–3 oz (57–85 g).

Habitat: Rocky foothills and canyon habitat within shrublands and open forests of piñon-juniper and mixed conifers.

Distribution: Throughout southern Colorado and into northern Colorado along the Front Range corridor.

Field Notes: The Colorado chipmunk scurries around broken, rocky country in the foothills and canyons of southern Colorado, and north along the foothills of the Front Range. It eats seeds, pine nuts, berries, and other vegetation, and the occasional insect. It stores food for winter, which it eats when it rouses periodically from dormancy. Colorado chipmunks breed in spring, producing a single litter of two to six young. They excavate nest burrows, preferring to conceal the entrance between the roots of trees or among the rocks. Chipmunk species are difficult to tell apart. The Colorado chipmunk is larger than the least chipmunk, whose range it overlaps, and a little stockier and heavier than the Uinta chipmunk. Both Colorado and Uinta chipmunks are more deliberate in their movements and less frenetic than least chipmunks. The Colorado chipmunk is more brightly colored than the Uinta chipmunk, but size and color are difficult to discern when the two species aren't side-by-side.

Legal Status: Nongame.

Hopi Chipmunk
Neotamias rufus

Field ID: Somewhat paler than other chipmunks, the Hopi is orangish brown to buff. The stripes along the back are brownish (not black) or lacking, and there are pale stripes through the eyes. The long tail is fairly bushy.

Size: Head/body length 4⅛–5 inches (105–127 mm), tail 3¼–3¾ inches (83–95 mm), weight 1¾–2 oz (50–57 g).

Habitat: Canyons, cliffs, slickrock, piñon-juniper woodlands.

Distribution: Far western Colorado south of the Yampa River, along the Gunnison River to the Black Canyon, and along the Colorado River into Eagle County.

Field Notes: If you're in the canyon country of western Colorado, you're probably seeing the Hopi chipmunk. These medium-sized chipmunks inhabit cliffs and rocky terrain. They stay close to the dens and hiding places they have among the boulders. They eat mainly seeds, as well as flowers and a small amount of insects. Gathering food in their cheek pouches, they store it for winter. They become dormant like other chipmunks, but for a shorter period, from December through late February or early March. They mate when they emerge from dormancy, producing one litter of two to six young. During summer, they may rest in their burrows during the heat of the day. The Hopi chipmunk was formerly considered a subspecies of Colorado chipmunk.

Legal Status: Nongame.

Uinta Chipmunk
Neotamias umbrinus

Field ID: This large chipmunk is brownish gray with black stripes down the back alternating with pale stripes.

Size: Head/body length 4½–5 in (114–127 mm), tail 3½–4⅗ inches (89–117 mm), weight 2–3 oz (57–85 g).

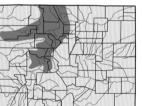

Habitat: Rocky, broken terrain, talus slopes, piñon-juniper woodlands, shrublands, high elevation pine forests, willow thickets.

Distribution: North-central and north-western Colorado.

Field Notes: You see a chipmunk, pull out this book, and try to tell which one it is. If you're in north-central Colorado, where the ranges of Uinta, least and Colorado chipmunks overlap, you may have a challenge. The least is the smallest and most frantic in its movements. The Uinta and the Colorado are similar in size and appearance and are difficult to tell apart. They both also like rocky terrain in pine and mixed-conifer woodlands. The Uinta inhabits sunny openings in heavy subalpine forest and is sometimes found on the alpine tundra. Its coloring isn't usually as bright as the Colorado chipmunk, and it isn't as sturdy-bodied. Like its cousins, the Uinta chipmunk feeds on berries, nuts, conifer seeds, flowers, other vegetation, and some insects. It excavates a burrow under rocks, fallen trees, shrubs, or tree roots and is a good climber. It becomes dormant in the winter months, rousing occasionally to feed on food it has cached. It breeds after emerging from dormancy, producing one litter of three to five young.

94 **Legal Status:** Nongame.

Yellow-bellied Marmot
Marmota flaviventris

Field ID: A large, plump-bodied ground squirrel, yellow-brown to red-brown, often with a whitish wash due to white-tipped guard hairs. The undersides are pale buff or yellowish and the tail is short and bushy.

Size: Head/body length 14–19 inches (36–48 cm), tail 4½–9 inches (11–23 cm), weight 5–10 lbs (2.2–4.5 kg).

Habitat: High mountain meadows, alpine tundra, rocky, open areas amid mountain forests.

Distribution: Mostly at higher elevations from the Eastern Slope westward across the state up to 14,000 feet.

Field Notes: Many hikers have reached the top of one of Colorado's 14,000-foot peaks to be greeted by yellow-bellied marmots. These year-round, mountaintop residents survive winter in the most severe habitat in the state by hibernating in burrows dug beneath rocks or other cover. Depending on conditions, some animals may stay in this winter sleep for as long as eight months. Marmots are one of the most visible high elevation animals. Hikers often hear the high-pitched calls of these plump rodents, nicknamed "whistle pigs," before they see them trundling around the rocks. Marmots are social animals, with colonies consisting of a dominant male with several females and their young. They feed on wildflowers and soft plant parts, adding seeds in late summer before they reenter their dens to sleep away the winter. They mate soon after emerging from hibernation, each female producing a single litter of three to eight young. The young emerge and are weaned by midsummer, and before long they all go underground again to hibernate. Between months spent in hibernation and time spent sleeping or sheltering, marmots may spend as much as 80 percent of their lives underground.

Legal Status: Small game.

White-tailed Antelope Squirrel
Ammospermophilus leucurus

Field ID: This slender ground squirrel is reddish brown to gray, with a single, white stripe along each side from shoulder to hip. The tail is medium-long and slightly bushy.

Size: Head/body length 5½–6½ inches (14–16.5 cm), tail 2–3 inches (51–76 mm), weight 3–5½ oz (85–156 g).

Habitat: Western shrublands, piñon-juniper woodlands, low-elevation riparian areas

Distribution: Far western Colorado up to about 7,000 feet.

Field Notes: This busy ground squirrel of the Southwest dashes frantically around gathering food or darting in and out of cover. Its name derives from the white flash of the underside of its tail, which it curls up over its back when it runs. It also reminded early settlers of the white tail flash of a fleeing pronghorn (antelope). It superficially resembles a golden-mantled ground squirrel but has only two pale stripes that lack black edges. It keeps numerous escape burrows as refuges from danger and extreme summer heat. Home burrows, which it digs or adopts from a kangaroo rat or other animal, are deeper and hold caches of food and a nest chamber. Adapted for life in a dry land, antelope squirrels don't need to drink free water but derive moisture from their food, which includes wildflowers, yucca, prickly pear, fruits, berries, insects, and occasional smaller mammals such as pocket mice. Insect remains sometimes mark the entrance to their burrows. They breed from February to April, producing a single litter of 5 to 11 young. In extreme heat, they retreat into their burrows and may cool their bodies by crawling with their scantily haired bellies pressed to the cooler, bare soil. Colorado is at the far northeastern edge of this squirrel's range.

Legal Status: Nongame.

Wyoming Ground Squirrel
Spermophilus elegans

Field ID: This medium-sized ground squirrel is pale gray with a medium-long tail bordered in white. It lacks the distinctive stripes or markings of other ground squirrels and chipmunks.

Size: Head/body length 7¼–9½ inches (18–24 cm), tail 2–4½ inches (51–114 mm), weight 10¾–17½ oz (305–496 g).

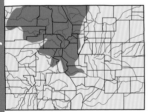

Habitat: Shrublands, grasslands.

Distribution: North central and north-western Colorado between 6,000 and 12,000 feet

Field Notes: Wyoming ground squirrels are entertaining to watch, as they scurry busily about, chase each other, dart down burrows, flick their tails, or pop upright in an alert posture. This habit led cowboys, miners, and settlers to nickname them "picket pin" for the stakes hammered into the ground to picket horses in open country. Wyoming ground squirrels live in groups but without the complex social relationships of prairie dogs, for which they are sometimes mistaken. They excavate complex burrows that may have up to eight entrances, which sometimes have dirt mounds to the side of the entrance. Wyoming ground squirrels use many vocalizations, with different calls for alarm, all clear, and greeting. They feed on grasses, flowers, and soft vegetation during the warm months. Some animals enter their hibernation dens by late July. They emerge from hibernation by late March or early April and soon mate, producing one litter of 3 to 11 young.

Legal Status: Game species.

Golden-mantled Ground Squirrel
Spermophilus lateralis

Field ID: This medium-sized, pale brown to golden ground squirrel has a red-brown mantle over the head and shoulders. Each side of the back has a dark stripe next to a white one. There is a white eye-ring but no white stripes on the face. The medium-length tail is hairy but not bushy.

Size: Head/body length 6–8 inches (15–20 cm), tail 2½–4¾ inches (64–121 mm), weight 6–9¾ oz (170–276 g).

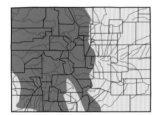

Habitat: Rocky slopes and open woodlands in foothills and mountains.

Distribution: Throughout the foothills, mountains, and mesas of central and western Colorado.

Field Notes: The golden-mantled ground squirrel is among the most familiar of mountain animals, scurrying around rocky slopes, turning up to beg at summer picnics and charming everyone with its cuteness and antics. It is often mistaken for a big chipmunk. The giveaway is the lack of stripes on the head. The ground squirrel's coloring is also more faded than the least chipmunk, its most likely neighbor. It eats seeds, wildflowers, green plants, fungi, and sometimes bird eggs and nestlings. Its den is hidden beneath a boulder or vegetation on a rocky slope, usually in or adjacent to open mountain forests. Golden-mantled ground squirrels are primarily solitary and are territorial about their dens. Once the summer tourist season ends, so does the ground squirrel's busy time, and it enters its den to hibernate through winter. Males emerge first in spring, females a week or so later. They soon breed and produce a single litter of two to eight young.

Legal Status: Nongame.

Sleeping through the Season

Some small mammals such as bats and ground squirrels survive winter by hibernating. The triggers that cue an animal to go into hibernation are still a mystery. They probably include shortening day length, dropping temperature, and scarcity of food.

A hibernating animal's metabolism slows down to a point that it appears to be dead. Heart rate, breathing, and temperature drop drastically. The animal's systems come almost to a standstill. It doesn't eat, drink, defecate, or urinate. This reduces the body's demands for food and water, which may be unavailable at that season.

The hibernator's body still needs some energy even for its minimal body processes, so it lives on stored body fat. Before entering hibernation, the animal eats voraciously to store enough fat to get it through until spring. It also must find a secure place to hibernate, called a hibernaculum, that is dry and maintains a steady temperature above freezing. Hibernation is a risky survival strategy. Every winter, a percentage of hibernating animals burn up their fat reserves too early and starve or freeze to death before spring.

Bears are the only large mammals that reduce some of their metabolic processes to survive winter, but this mechanism differs from the hibernation of small mammals. While a bear's breathing and heart rate drop, its body temperature does not go down significantly. It also can rouse fairly easily if disturbed. Because of this, biologists contend that bears are not "true hibernators." They are, however, doing something magical that allows them to survive winter without eating, drinking, defecating, or urinating, while also giving birth and nursing young. Perhaps we need a new term such as "ursine hibernation" to describe a bear's winter survival strategy.

Whatever we call these processes, the body changes bears and other hibernating animals are capable of are truly amazing.

Spotted Ground Squirrel
Spermophilus spilosoma

Field ID: This small, grayish brown squirrel has a faint, mottled pattern of pale, squarish spots on the back, a medium-length, nonbushy brown tail and pale undersides.

Size: Head/body length 5–6 inches (13–15 cm), tail 2¼–3½ inches (57–89 mm), weight 3–4½ oz (85–128 g).

Habitat: Arid grasslands with sparse vegetation and sandy soils.

Distribution: Eastern Plains and a small area of far southwestern Colorado in Montezuma County.

Field Notes: This somewhat nondescript ground squirrel is not as broadly distributed as the thirteen-lined ground squirrel. It inhabits semiarid grasslands, preferring sparse, even overgrazed, cover. One of the smallest ground squirrels in North America, it is scarcely larger than a chipmunk. As an inhabitant of open country with little cover, it is extremely shy and secretive. Its movement pattern is a rapid scurrying with abrupt stops as it checks for danger. The animal moves with its body and tail close to the ground, frequently darting down its burrow. The burrows have several entrances beneath shrubs or covering plants. Spotted ground squirrels feed on grass, wildflowers, green vegetation, and seeds, as well as occasional insects or animal material. They are diurnal, most active during the middle of the day, though they may retreat to their burrows when it is very hot. They hibernate during winter, entering their dens as early as late July and reemerging in early April—yearlings first, followed by adult males. Adult females emerge one to two weeks later. They breed and produce one litter of 5 to12 young.

Legal Status: Nongame.

Thirteen-lined Ground Squirrel
Spermophilus tridecemlineatus

Field ID: This ground squirrel is buffy to brown, with thirteen alternating dark and light stripes running down the back. The dark stripes have squarish pale spots. The tail is medium-long. The four-toed front paws have long, digging claws and the hind paws have five toes.

Size: Head/body length 4½–6½ inches (114–165 mm), tail 2½–5¼ inches (64–133 mm), weight 5–9 oz (142–255 g).

Habitat: Grasslands.

Distribution: Throughout eastern Colorado, San Luis Valley, South Park, far northwestern Colorado.

Field Notes: With their long, lanky bodies, patterned backs, and low-slung, loping gait with long tail held horizontal, the thirteen-lined ground squirrel sometimes looks like a mammalian snake undulating across open ground. It typically scurries along, pops suddenly upright to look for danger, then darts down its burrow. The burrows have a narrow, inconspicuous hole, usually concealed by vegetation and lacking telltale loose dirt. Thirteen-lined ground squirrels feed on grasses, forbs, fungi, and flowers. Half of their diet is animal matter, mostly insects but also lizards, snakes, birds, and even nestling rabbits. They are much smaller and shyer than prairie dogs. Preyed upon by many, they venture out only during the day when the sun is high and they can spot predators with their large eyes. They are generally solitary. These ground squirrels give a trill-like alarm call in response to disturbance or intruders. They enter hibernation late summer through fall, plugging their burrow entrances and curling into a ball, emerging again in early April or May. They breed soon after and produce six to nine young. In some parts of their range, females may have a second litter.

Legal Status: Small game.

Rock Squirrel
Spermophilus variegatus

Field ID: This large ground squirrel has a mottled gray to charcoal coat, often mixed with black or cinnamon. The tail is long and fairly bushy.

Size: Head/body length 10–11 inches (25–28 cm), tail 7–10 inches (18–25 cm), weight 1½–2 lbs (680–907 g).

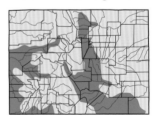

Habitat: Canyons, rocky hillsides and outcrops, particularly amid piñon-juniper woodlands and shrublands, also rocky areas in montane forest, riparian, urban, and suburban areas.

Distribution: Throughout central, south-central, and southwestern Colorado below about 8,300 feet.

Field Notes: Rock squirrels are the largest of Colorado's ground squirrels. Due to their size, shape, color, and fairly bushy tail, they are often mistaken for fox squirrels. They can often be spotted scampering over rocky terrain, undulating up and over rocks, trailing their furry tails. Like prairie dogs with bushy tails, they build elaborate burrow systems beneath boulders. And while they do live in colonies and whistle warnings to each other, they don't have the elaborate social structure that prairie dogs do. Colonies have a dominant female, with dominant males defending territories during the breeding season. Adult males and females have little to do with each other except during the spring breeding season. They breed from late spring through early summer, producing a single litter averaging five young. In the more northern parts of their range, they hibernate through winter. Though ground-dwellers, they are adept at climbing trees. They feed on berries, seeds, acorns, and fruits, as well as flowers and tender vegetation, and the occasional insect or carrion.

Legal Status: Small game.

Gunnison's Prairie Dog
Cynomys gunnisoni

Field ID: This plump prairie dog is tan with dark fur on the cheeks, brow, and top of the head. It has a short, gray-tipped tail.

Size: Head/body length 10½–15 inches (27–38 cm), tail 1½–2½ inches (38–64 mm), weight 1–3 lbs (454–1,360 g).

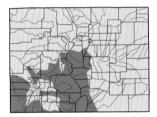

Habitat: Grassy, open areas of western shrublands and mountain parks.

Distribution: South-central and southwestern Colorado from 6,000 to 12,000 feet.

Field Notes: Gunnison's is the smallest of Colorado's three prairie dog species. It lives in colonies, but these are only loosely organized and the animals are much less social than black-tailed prairie dogs. They do, however, communicate with alarm barks and other calls. The burrows are scattered more haphazardly around the terrain and lack the elaborate mounds familiar from the extensive colonies of black-tailed prairie dogs. Gunnison's prairie dogs don't clip tall plants around their colony, and their burrows may be dug under rocks and other features. They are sometimes confused with Wyoming ground squirrels. Unlike the year-round-active black-tailed prairie dog, Gunnison's prairie dogs, which inhabit higher elevations, hibernate in winter. They mate soon after emerging, producing a single litter averaging three to four young. Gunnison's prairie dogs feed on grasses and sedges, as well as flowers, forbs, and shrubs. They don't usually dig for roots.

Legal Status: Nongame, a candidate for listing on the endangered species list.

White-tailed Prairie Dog
Cynomys leucurus

Field ID: This large prairie dog has a yellowish buff to whitish gray coat and white-tipped tail. There are distinctive dark brown to black patches on the head, extending from the cheeks to above the eye.

Size: Head/body length 10¾–13½ inches (27–34 cm), tail 1½–2½ inches (38–64 mm), weight 1½–3¾ lbs (680–1700 g).

Habitat: Shrublands, grasslands, mountain parks and valleys.

Distribution: Far northwestern and west-central Colorado, especially along the Gunnison and Colorado river drainages.

Field Notes: Restricted to the northwestern part of the state, the white-tailed prairie dog has the smallest range in Colorado of the three prairie dog species. Its colonies are more loosely organized than those of black-tailed prairie dogs, with less interaction between individuals, though they do respond to each other's alarm barks. Those familiar with the characteristic mounds and burrows of a black-tailed colony might not recognize the haphazard scattered holes of a white-tailed community, many of them beneath a rock or shrub and lacking dirt mounds around the entrance. Underground, though, the burrows are extensive and elaborate, with up to 100 feet of tunnels and different chambers for sleeping, hibernating, and raising babies. White-tails feed on grasses, forbs, and shrubs. Some populations hibernate in winter. They mate soon after emerging, producing a single litter averaging five young. The first restorations of endangered black-footed ferrets in Colorado were into white-tailed prairie dog colonies near Dinosaur National Monument.

Legal Status: Small game.

Black-tailed Prairie Dog
Cynomys ludovicianus

Field ID: This large, plump prairie dog is tan with a short, black-tipped tail.

Size: Head/body length 11–13 inches (28–33 cm), tail 3–4 inches (76–102 mm), weight 2–3 lbs (900–1,360 g).

Habitat: Eastern Plains and foothills grasslands, urban/suburban open space, and undeveloped lots along the Front Range corridor.

Distribution: Eastern Colorado from Front Range foothills to the Kansas state line.

Field Notes: The black-tailed is the most abundant, widespread, and familiar prairie dog in Colorado. These highly social animals live in large colonies, with elevated mounds around their burrow entrances that allow sentinels a sight perch and keep water from running into the burrow. Black-tailed prairie dogs produce one litter of four to five young a year. Young are born in late March and early April and emerge above ground in late May and early June, small carbon copies of the adults. Vocal communication is complex, with different calls for different predators and for the all clear. Black-tailed prairie dogs are active throughout the year, remaining underground for periods of inactivity during bad winter weather or on cloudy days. Their eyes are set toward the top of the skull to aid in spotting predators that might come from the sky as well as the ground. They feed on the leaves, flowers, seeds, shoots, and roots of grasses and forbs. Bison, pronghorn, and domestic cattle prefer to graze within prairie dog towns rather than other rangeland because of the higher protein and nitrogen content of the vegetation.

Prairie dogs have been the subject of intense eradication efforts since the late nineteenth century. They are still classified under Colorado agriculture statutes as an agricultural pest. Conversion of their grassland habitat to agriculture, rangeland, and urban/suburban development has reduced this species to an estimated 2–5 percent of its original range. Sylvatic plague, a nonnative disease, has also decimated prairie dog populations. However, in 2009, the US Fish and Wildlife Service determined the black-tailed prairie dog did not warrant protection as a threatened or endangered species. Population declines in prairie dogs led to the near-extinction of the black-footed ferret, which depends upon prairie dogs for food and shelter.

Legal Status: Species of special concern.

It Takes a Village

A prairie dog town is much more than a bunch of animals milling around. It's a highly organized "village" of multiple families, each consisting of a male, several females, and their young. Each family occupies a burrow and plays, grooms each other, shares food, and maintains the burrow. They communicate using a sophisticated repertoire of calls, with different sounds for danger, pleasure, anger, and the all clear.

Group living offers good protection from predators. While family groups feed, sentinels keep watch for danger, sitting upright atop their burrows. If a sentinel sees a coyote, hawk, or even a human, they alert the group with a series of alarm barks. The call is taken up by the community, and soon half of the animals may begin barking. When the danger is gone, the sentries sound an all clear, tossing back their heads and upper bodies in what's called a jump-yip.

This effective alarm network offers protection for other species as well. Burrowing owls build their nests in abandoned prairie dog burrows, taking advantage of the watchfulness of their hosts for their own defense.

Abert's Squirrel
Sciurus aberti

Field ID: Most Abert's squirrels along the Front Range are jet black. Some are gray with a white underside and black stripes on the sides. Those in southwestern Colorado are salt-and-pepper gray. Some squirrels are brown. All have dramatically tufted ears and bushy tails.

Size: Head/body length 11–12 inches (28–30 cm), tail 8–9 inches (20–23 cm), weight 1½–2 lbs (680–900 g).

Habitat: Ponderosa pine forest.

Distribution: Foothills and mountains of central and southern Colorado.

Field Notes: Abert's squirrel is also called the tassel-eared squirrel for the tufts of fur on its ears. These handsome squirrels depend on ponderosa pines for shelter and food, eating the seeds, cones, twigs, bark, and buds. A pile of broken pine cones, gnawed twigs, and clipped-off clusters of needles beneath a ponderosa pine is a sure indicator of their presence. They are usually solitary, but in spring groups of males follow one female around, scurrying across the forest floor and leaping up and down trees, competing for her attention. They mate in April or May, producing one litter of two to four young. They are active year-round, using nests of pine twigs set in a pine tree and lined with grass, or they may burrow into a mass of dwarf mistletoe in the branches of a pine tree. In contrast to the small but feisty chickaree (pine squirrel), Abert's squirrels are not territorial and rarely chatter at visitors entering their forest. The lucky hiker is more likely to just catch a glimpse of one of these large, handsome squirrels as it leaps from forest floor to tree trunk. Because their life history is governed by the pine nut crop, their numbers and distribution vary.

Legal Status: Game species.

Fox Squirrel
Sciurus niger

Field ID: This large squirrel has a reddish brown or gray-brown coat, with reddish, golden, or whitish underparts. Its long, bushy tail is usually curled over the back. It has small ears, large eyes, and dexterous paws that can hold and manipulate food.

Size: Head/body length 10–15 inches (25–38 cm), tail 9–14 inches (23–36 cm), weight 1⅕–3 lbs (544–1361 g).

Habitat: Wooded areas, parks, and backyards in cities and towns, woodlands along streams and rivers, trees around farms and ranches.

Distribution: Eastern Colorado into the foothills of the Front Range.

Field Notes: The familiar fox squirrel of backyards, parks, and neighborhoods migrated west into Colorado as settlers planted trees across the Great Plains. They were also intentionally released in Greeley, Denver, and other towns. Trees are key to their life history. They eat nuts, chew on tree bark and buds, climb trees for safety, inhabit tree holes, and build their nests—messy collections of leaves set high in the branches—in trees to raise their babies and to stay warm in winter. Fox squirrels don't hibernate, but when it's very cold, they curl up in their nests and sleep. Fox squirrels are very active and visible, their running, climbing, chasing, chattering, and bird-feeder-robbing either entertaining or annoying, depending on the viewer. The genus name, *Sciurus*, means "shade tail," a reference to how the squirrel's bushy tail often curls up and over its head. Waving of the fluffy tail may challenge another squirrel, telegraph excitement, or try to flush a predator from hiding. Fox squirrels also communicate by chattering, barking, and scolding. Mature females raise two litters a year, averaging three to four young.

Legal Status: Small game.

Pine Squirrel (Chickaree)
Tamiasciurus hudsonicus

Field ID: This small tree squirrel is gray with paler undersides, dark stripes on the sides, and a long, furry tail.

Size: Head/body length 7–8 inches (18–20 cm), tail 4–6 inches (102–152 mm), weight 7–9 oz (198–255 g).

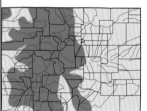

Habitat: Mountain coniferous forests.

Distribution: Foothills and mountains of central and western Colorado.

Field Notes: Hikers through high mountain forests have likely made the acquaintance of this feisty squirrel, the smallest of Colorado's tree squirrels. Highly territorial, it scolds invaders with a high-pitched chattering that has led to its nickname, chickaree. The pine squirrel's principal food is conifer seeds, but it also eats buds, berries, leaves mushrooms, and other fungi, It spends much of its time harvesting the cones of conifers and caching them in enormous middens— mounds of cones, nibbled cores, and cone scales—beneath spruce and lodgepole pine trees in mountain forests. The squirrels scamper about the tree branches cutting cones. Hikers sometimes hear the cones hitting the ground. A busy squirrel can harvest a cone every two to three seconds. Squirrels also perch on branches and nibble on cones, the debris raining down on the midden below. Pine squirrels are active year-round but stay in their nests during very cold weather. Nests are made of twigs and leaves and set high in trees protected within a dense forest canopy. Pine squirrels breed from April through June, producing a single litter of two to five young. In other parts of the United States, the pine squirrel is known as the red squirrel.

Legal Status: Small game.

Speaking Squirrel

Anyone who has ever been scolded by a squirrel in their backyard, or walked past a prairie dog town and triggered a frenzy of alarm barks, knows that squirrels (tree or ground) can be a very vocal bunch.

The barks, chatters, whistles, and "chucks" made by squirrels are not just random noise. They have different meanings and comprise a complex system of vocal communication. A University of Kansas study of thirteen-lined ground squirrels identified six distinct calls. A *trill* was an alert or warning. A *long purr* was an alert. A shorter *purr* signaled a disturbance nearby. A *squeak* meant the animal was distressed. *Tooth chattering* was a warning threat, as was a *growl*. Likewise, prairie dogs "talk" to each other with various social calls and have different calls for avian, four-legged and two-legged intruders, and another for the all clear.

Pine squirrels (chickarees) also differentiate between intruders on the ground versus the air. They *chuk-chuk* loudly at humans, foxes, and terrestrial threats. But when they detect a hawk, their alarm call is a short, high-pitched *seet*, similar to the alarms raised by songbirds in reaction to a raptor.

Marmots use a variety of whistles as alarms, warnings, and threats, which their neighbors, pikas and golden-mantled ground squirrels, also react to.

Sit and observe ground squirrels for a while and you will start to understand the meaning of some of their many calls.

Family Geomyidae—Pocket Gophers

In contrast to the diverse squirrel family, pocket gophers are a small family with little difference among the species. They are all fossorial, living most of their life in underground tunnels. They have fur-lined, exterior cheek pouches (pockets) used for carrying seeds and food, not for holding excavated dirt, as some people suppose. Pocket gophers are highly adapted to a life below ground, with large, strong front feet and long, sharp, sturdy, curved claws for digging; a rounded head-body shape; and short fur that doesn't collect dirt. The lips close behind the teeth to keep out dirt. The eyes and ears are small, to reduce the risk of being filled with dirt. Sight is not an important sense for a subterranean dweller, but sound can be. Though the pocket gopher's external ears are minute, their auditory bullae, the hollow bulges at the base of the skull that contain the middle and inner ears, are very large. Gophers rely heavily on highly sensitive whiskers to gather information in their pitch-black world. Pocket gophers are rarely seen above ground but occasionally one can be spotted when it comes to the surface to push excavated dirt out of a burrow. Mounds of loose soil dot the ground in an area where pocket gophers are active. The startling sight of a flower pulled suddenly into the ground by its stem, seemingly by some invisible force, is evidence of a busy gopher underground, tugging tasty plant stems from below.

Control of gophers as pests has reduced many populations, and some populations of Botta's and northern pocket gophers are classified by Colorado Parks and Wildlife as species of special concern.

Animal Sign: Mounds of loose dirt with no obvious entrance; in winter, castings or eskers of excavated dirt hidden below the snow and revealed after snowmelt; dirt spraying up from a mound, indicating a gopher in the process of ejecting excavated dirt.

Botta's Pocket Gopher
Thomomys bottae

Field ID: This small gopher ranges in color from yellowish to red-brown. It has a large head; tiny eyes and ears; short, bare tail; long, curving claws on its front paws; and external, fur-lined cheek pouches.

Size: Head/body length 5–7 inches (13–18 cm), tail 2–3¾ inches (51–95 mm), weight 2½–9 oz (71–255 g).

Habitat: Grasslands, piñon-juniper woodlands, open foothills and mountain forests, shrublands, agricultural land, valley bottoms.

Distribution: Southwestern Colorado and south central Colorado.

Field Notes: Also called the valley pocket gopher, this animal lives in riparian areas and valley bottoms. It prefers soft, sandy soils found along creeks and rivers, where it feeds on roots, seeds, grasses, and forbs. Food is stored in the burrow, and the species practices coprophagy. Throughout its range, Botta's gopher varies greatly in size and coat color. Western Colorado animals are yellowish while those in the southeast part of the state are reddish. They are solitary animals except for a female with young. A population of Botta's gophers has about three adult females to every adult male. They breed between March and July and produce one litter of two to five young. Young disperse from the home burrow in late summer or fall. This is when these subterranean dwellers, without a burrow of their own, are most vulnerable to being picked off by hawks or coyotes.

Legal Status: Small game; some populations are species of special concern.

Northern Pocket Gopher
Thomomys talpoides

Field ID: This small gopher ranges in color from grayish yellow to dark brown. It has a large head; tiny eyes and ears; short, bare tail; long, curving claws on its front paws; and external, fur-lined cheek pouches. The orangish front teeth are in front of the lips.

Size: Head/body length 5–6½ inches (13–16.5 cm), tail 1¾–3 inches (44–76 mm), weight 2¾–4³/₅ oz (78–130 g).

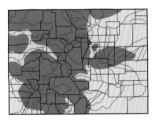

Habitat: Grasslands, shrublands, farmland, cattle pastures, alpine tundra, open areas amid woodlands.

Distribution: From the eastern foothills west across the state above 5,000 feet.

Field Notes: Northern pocket gophers live in more habitats, and thus are found in more of the state, than Colorado's other gopher species. They dig elaborate burrow systems as much as a foot-and-a-half belowground and feed on green vegetation, especially composites and legumes. In winter, roots are important food. These pocket gophers breed in late spring and early summer, producing one litter of four to six young. Though abundant, these subterranean dwellers are rarely seen, although evidence of their digging is obvious. They push the loosened soil to the surface, leaving it in a neat mound, then retreat into the burrow and plug the hole with more dirt. A small rodent that spends most of its time tunneling underground may seem insignificant, but pocket gophers have a major ecological impact on soils. Their tunneling and earth-moving aerates, mixes, and reinvigorates the soil, moving organic matter from the surface into deeper soils and exposing and redistributing deep layers. Scientists estimate a colony of gophers could move more than 400 tons of soil per hectare (10,000 square meters) in a year.

Legal Status: Small game; some populations are species of special concern.

Plains Pocket Gopher
Geomys bursarius

Field ID: This large gopher is pale to yellowish brown, with a large head; tiny eyes and ears; short, bare tail; long, curving claws on its front paws; and external, fur-lined cheek pouches.

Size: Head/body length 5½–9 inches (14–23 cm), tail 2–4½ inches (51–114 mm), weight 4½–12½ oz (128–354 g).

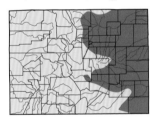

Habitat: Grasslands, farm and range-lands, roadsides.

Distribution: Throughout eastern Colorado.

Field Notes: The plains pocket gopher is distributed throughout the Eastern Plains, where the land is dotted with its earthen mounds. Its front teeth angle forward to aid in gnawing through the soil as it burrows. It feeds heavily on the underground parts of plants, especially grasses, as well as cactus and forbs. Plains pocket gophers breed in late winter and spring, producing one litter of one to six young per year. The burrow systems of gophers are quite elaborate and can reach 300 feet in length. Deep tunnels, below the soil's frost line, connect to shallower tunnels where the animals feed on roots and from which they can emerge above ground. There are side chambers for food storage, nesting, and waste. A study in Las Animas County found that the burrow systems of individual gophers overlapped, but at different depths, like a network of highway cloverleaves and flyovers, all carrying gophers this way and that underground without running into each other.

Legal Status: Small game.

Yellow-faced Pocket Gopher
Cratogeomys castanops

Field ID: This large, reddish brown gopher has a wash of yellow on cheeks and neck. It has a large head; tiny eyes and ears; short, bare tail; long, curving claws on its front paws; and external, fur-lined cheek pouches.

Size: Head/body length 7¼–8 inches (18–20 cm), tail 3–4 inches (76–102 mm), weight 7½–11¾ oz (213–333 g).

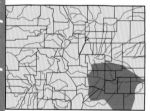

Habitat: Grasslands, shrublands, farmland.

Distribution: Southeastern Colorado.

Field Notes: The yellow-faced pocket gopher prefers deep, sandy soils. It will leave mounds of soil, but often the evidence of its activity is a plug of dirt that settles and sinks below ground level. Its range extends from Colorado south into Mexico, and it was formerly known as the Mexican pocket gopher. It breeds from late March through August, producing up to three litters of one to five young. Adult females may outnumber adult males by four to one. They feed on green plants and roots. Pocket gophers sometimes forage for food out in the open but usually eat roots and tubers and pull plants down into their burrows by the roots, a very odd thing to witness from the surface. When they do emerge into the upper world, they don't venture far from their holes because their small eyes are not good tools for detecting predators. Leading a life underground means they don't have to alter their behavior for the changing seasons and thus are active year-round.

Legal Status: Small game.

Family Heteromyidae—Pocket Mice and Kangaroo Rats

Like pocket gophers, heteromyids have external, fur-lined cheek pouches for carrying food. They have large heads with little definition of the neck, large eyes, and small external ears.

These animals have short front legs and long to very-long hind legs, and they move by jumping. Pocket mice leap on four legs (sometimes two), but kangaroo rats are adapted for an extreme ricochet style of locomotion, leaping on their hind legs and turning in midleap with a flick of the tail.

Adapted for life in an arid climate, pocket mice and kangaroo rats do not need to drink free water. They are nocturnal, staying in their burrows by day and emerging to forage at night.

Animal Sign: Hopping track showing only two feet, with a tail dragline.

Olive-backed Pocket Mouse
Perognathus fasciatus

Field ID: This small mouse's coat is buff to greenish brown intermixed with black hairs in a darker panel down the center of the back, a panel of yellow on each side, yellow patches behind the ears, and a white belly. It has long hind feet and a long tail.

Size: Head/body length 2⅘ inches (71 mm), tail 2½ inches (64 mm), weight ¼–⅓ oz (7–9 g).

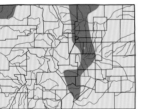

Habitat: Grasslands, shrublands.

Distribution: Along the foot of the Front Range corridor from the Wyoming state line to Las Animas County and in the extreme northwestern corner of the state.

Field Notes: Olive-backed pocket mice are quite small, their bodies only 2 to 3 inches long, trailing a tail as long or longer. They feed on grass and forb seeds, some green vegetation, and occasional insects. Like pocket gophers, pocket mice have external, fur-lined cheek pouches they use to carry seeds back to their burrows. Using their paws, they stuff food into the pouches, return home, then paw at the pouches to dump out the seeds. The olive-backed pocket mouse prefers sandy soils, in which it digs burrows beneath the cover of yuccas, rabbitbrush, and clumps of grass. It is known to sand bathe, rubbing its coat and fur-lined cheek pouches in the sand to clean its fur. The olive-backed pocket mouse has one or two litters a year of four to six young. It was formerly known as the Wyoming pocket mouse because its range is primarily across the northern Great Plains. Its range in Colorado is limited to a narrow finger extending southward along the foot of the Front Range corridor.

Legal Status: Nongame.

Plains Pocket Mouse
Perognathus flavescens

Field ID: This mouse ranges in color from buff to yellow to reddish intermixed with black hairs, with a white belly and often a buff patch behind the ear. It has long hind feet and a long tail.

Size: Head/body length 2¼–2¾ inches (57–70 mm), tail 2–2¾ inches (51–70 mm), weight ¼–⅓ oz (7–9 g).

Habitat: Grasslands.

Distribution: Throughout eastern Colorado, the San Luis Valley, and in parts of far western Colorado.

Field Notes: The plains pocket mouse deserves its name, being the most common pocket mouse of the Eastern Plains. It prefers areas with sandy soil. Though it lives in prairies and rangelands, it will also make its home in habitat that has scanty cover and bare ground. Its shallow burrows are often dug into a pocket gopher mound or may be hidden beneath a yucca or prickly pear. When in the burrow, the mouse plugs the opening with soil and may place another plug further along in the burrow. Plains pocket mice eat the seeds of grasses, weeds, wildflowers, and other forbs. Plains pocket mice trapped in grain fields were found to have weed seeds, rather than grains, in their cheek pouches. Using their external cheek pouches, they carry seeds back to their den to cache for winter. They become dormant in cold weather, rousing to feed on their stored food. They breed May through July, producing one or two litters of three to four young.

Legal Status: Nongame.

Silky Pocket Mouse
Perognathus flavus

Field ID: The coat of this small, pale pocket mouse is buff to pink-ish brown mixed with black hairs. It has a large patch of buff fur behind the ear.

Size: Head/body length 2–2½ inches (51–64 mm), tail 1¾–2¼ inches (44–57 mm), weight ¼–⅓ oz (7–9 g).

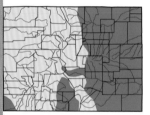

Habitat: Grasslands, piñon-juniper woodlands.

Distribution: Throughout eastern Colorado, west along the Arkansas River to Salida, San Luis Valley, far southwestern Colorado.

Field Notes: This pocket mouse is well named: its coat is especially soft and silky. That, along with its small size and the large pale patch behind the ear, make it easy to distinguish from other pocket mice. The silky pocket mouse feeds on grass, weed, wildflower, and other forb seeds. Its burrow is hidden beneath a yucca, cactus, shrub, or other protective vegetation and it rarely ventures far from the burrow entrance. Unlike the deep, elaborate burrows of some fossorial rodents, this mouse's burrow is shallow, only about four inches deep, and constructed within a dirt mound with several entrances. Inside the mound is a chamber, with a second, larger chamber down another tunnel below it. The burrows have side chambers for food storage and waste. The mouse caches seeds in several chambers and will keep caching seeds even when the existing caches contain abundant food. Like other pocket mice, the silky pocket mouse becomes dormant in winter, depending on the weather. It also does not need to drink free water but manufactures water metabolically. It produces one litter a year of two to six young, and occasionally a second litter.

Legal Status: Nongame.

Great Basin Pocket Mouse
Perognathus parvus

Field ID: This medium-sized pocket mouse is sandy-yellow to olive gray with pale underparts, pale patches behind the ears, and a long tail that is dark on top, light underneath, and has a tuft at the tip.

Size: Head/body length 2½–3 inches (64–76 mm), tail 3¼–4 inches (82–102 mm), weight ⅔–1 oz (19–28 g).

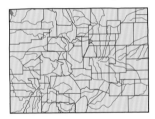

Habitat: Shrublands, grasslands, piñon-juniper woodlands.

Distribution: Browns Park in far northwestern Colorado.

Field Notes: This long-tailed pocket mouse is known in Colorado only from a handful of specimens collected in Browns Park. It is quite common, however, in the desert Great Basin of Nevada, Utah, Idaho, and eastern Washington and Oregon, where it may be the most abundant mammal. Great Basin pocket mice live in burrows that they plug with dirt to keep out cold, heat, and predators. They survive winter by becoming dormant and feeding on stored food when they rouse. Like other pocket mice, they have fur-lined external cheek pouches and feed on grass, weed, wildflower, and other forb seeds, some green plants, and the occasional insect. They breed after emerging in the spring and have one, sometimes two, litters of three to four young.

Legal Status: Nongame.

Hispid Pocket Mouse
Chaetodipus hispidus

Field ID: This large pocket mouse has a bristly coat that is pale yellowish to reddish intermixed with black hairs. It has buff sides that grade into a white belly. The long, haired tail is dark down the top, pale underneath, and shorter than the head and body.

Size: Head/body length 4½–5 inches (114–127 mm), tail 3½–4½ inches (89–114 mm), weight 1–1⅔ oz (28–47 g).

Habitat: Grasslands, dry riparian areas, weedy fields, and roadsides.

Distribution: Throughout eastern Colorado.

Field Notes: The hispid is the largest pocket mouse in the state. While most pocket mice have silky hair, the hispid's coat is coarse and feels rough when stroked from rump to head (*hispid* means "bristly"). It is a solitary creature, moving around the prairie at night gathering the seeds of grasses, wildflowers, and sagebrush. It often shares habitat with the more numerous plains pocket mouse. Pocket mice become dormant during winter, retreating to their dens but rousing occasionally to feed on stored food. They build their burrows beneath a yucca, rabbitbrush, or other shrub. There may be several entrances, which the mouse plugs during the day to keep out heat and cold and to deter predators. The burrow entrance is just wider than the mouse's head. Unlike other pocket mice, the hispid leaves small piles of soil outside its burrows, like a mini version of a pocket gopher mound. This species breeds in spring and summer and may have two or more litters of five or six young, which leave the nest by the time they are a month old.

Legal Status: Nongame.

Ord's Kangaroo Rat
Dipodomys ordii

Field ID: The coat of this beautiful native rat is yellowish brown intermixed with black hairs, with white underparts, legs, and underside of tail. There is a white pattern on the face. The hind legs and feet are very long and the front legs are quite small. The very long, haired tail has a dark stripe down the center, a fur tuft at the tip, and is longer than the head and body.

Size: Head/body length 4–4½ inches (102–114 mm), tail 5–6 inches (13–15 cm), weight 1½–2½ oz (43–71 g).

Habitat: Shrublands, grasslands, piñon-juniper woodlands.

Distribution: Throughout eastern Colorado from the Front Range corridor to Kansas, the San Luis Valley, far northwestern Colorado, far southwestern Colorado.

Field Notes: The kangaroo rat is well named. Its long, strong hind legs and feet, small front legs, and habit of boinging around on two legs like a pogo stick make it look like a tiny, rodent kangaroo. It even has external pouches, but they are cheek pouches for carrying food, not a belly pouch for carrying young. Kangaroo rats feed on grass, forb, and weed seeds, and the occasional insect. They are well adapted to life in arid landscapes and don't require free water (their bodies produce it metabolically from their food). In times of drought and extreme hot weather, they estivate—the hot-season version of hibernating. They plug their burrow entrance to keep out heat and reduce moisture evaporation from their bodies and breathing. By contrast, they don't hibernate but are active all winter. Breeding is greatly affected by precipitation and the emergence of green vegetation. They may breed any time from February through August, producing litters of two to four young. If a female breeds early enough, she may produce a second litter later in the summer. Kangaroo rats are nocturnal and may be seen jumping around by the light of the moon, though they are less active at the full moon, when predators on the prowl could more easily see them.

Legal Status: Nongame.

Family Castoridae—Beavers

This family has only one living genus, *Castor*, with only two species. The European and American beavers are similar in appearance and lifestyle. While the American beaver may be the single member of its family in North America, it is among the most identifiable and, to some people, charming of animals. The beaver's tree-cutting and damming, however, make it a nuisance to many landowners and communities.

Beavers are highly adapted to an aquatic lifestyle and are valued for their rich furs—thousands are trapped annually in Colorado. The popularity of beaver-fur hats in Europe in the nineteenth century led to intensive trapping that nearly eradicated beavers from the West, but populations have rebounded, and beavers, which are highly mobile, regularly colonize stretches of waterways statewide.

Animal sign: Streamside tree stumps chiseled to points, with grooved teeth marks; woodchip debris around chiseled trees; standing trees with deeply chiseled divots half the diameter of the trunk; logs and downed branches with chiseled ends; peeled, barkless logs and twigs; dams of mud and sticks; large, domed earth-and-stick lodge in a pond; slick mud slides at water's edge; tracks at water's edge—webbed, five-toed hind track; handlike, five-toed front track.

American Beaver
Castor canadensis

Field ID: These largest of North American rodents have thick, dark-brown coats, rounded heads, short ears, blunt noses, and wide, flat tails.

Size: Head/body length 2–2½ feet (61–76 cm), tail 9–10 inches (23–25 cm), weight 30–70 lbs (14–32 kg).

Habitat: Mountain streams, prairie creeks, ponds, canals, reservoirs.

Distribution: Statewide.

Field Notes: Many people think of beavers as mountain animals, but they will move into any stretch of stream from prairies to peaks to western canyons, as long as there is woody vegetation. They are marvelously adapted for an aquatic lifestyle, with webbed hind feet for paddling and a wide, flat tail that acts as a rudder. Prior to diving, beavers can pinch shut their ears and nostrils and cover their eyes with a special membrane, a clear, inner eyelid that acts like natural goggles to protect the open eyes underwater. Beavers feed on the bark, buds, leaves, and twigs of trees such as willows, cottonwoods, and aspen, as well as other green vegetation. Microorganisms in the gut help beavers digest wood. Those unique flat tails are famous as a means of alerting other beavers of danger. One sharp slap on the water surface by a sentinel and other beavers know to dive to safety. The tail is also a storehouse for fat and helps the animal balance when it sits upright to chew. Beavers are social animals that live in family or extended-family groups of bonded adults with their young and young from previous years. Families number from four to eight animals. Beavers build dams of sticks and mud across streams. When the water backs up to form a pond, they construct a mud-and-stick lodge out in the water as a refuge from predators. The entrance is below water, but the interior of the lodge is above the water line. Beavers living along a stream build dens in the bank. Beavers don't hibernate but stay in their lodges through harsh weather, feeding on twigs and branches cached below the water. The location of the lodge in the middle of a frozen pond, with an underwater entrance, allows these large rodents to come and go beneath the ice, safe from predators. Beavers mate in winter and have a single litter of two to five young. Litter sizes are smaller at higher elevations. Young beavers disperse when they are two years old, and the family will abandon a pond once the food supply is exhausted, colonizing a new stretch of stream and starting again the process of dam building and pond succession.

Legal Status: Furbearer.

Nature's (Eco) Engineers

Humans are fascinated by beavers, calling them nature's engineers and referring to being *busy as a beaver.* We might broaden that nickname to *nature's eco-engineers.* The activity of beavers has a major impact on natural communities. A beaver family can fell numerous trees along a stream or pond in a short time. Their damming and tree-cutting contribute to plant succession—the dams cause flooding, which kills additional trees and creates a pond or wetland. This also encourages growth of willows, a food favored by beavers. The pond eventually silts in and is invaded by grasses and forbs, resulting in a meadow that is eventually invaded by trees again. The dams, ponds, and wetlands that beavers create raise the water table and slow down spring runoff. Their actions create habitat for all sorts of plants and animals and keep natural communities in a constant state of change.

Family Cricetidae—Native Mice, Rats, and Voles

This diverse and abundant family includes the deer mice, harvest mice, woodrats, and voles. They tend to be small to medium-sized, ranging from the ⅕-ounce plains harvest mouse to the 4-pound muskrat. Most are nocturnal. Many move about furtively at the soil surface, hidden by grass and vegetation. The presence of woodrats, though, is never in doubt. Though the animals themselves are seldom seen, their large, messy "houses"—collections of sticks and found (or stolen) items—are impossible to miss.

Extremely important as a prey base, these abundant rodents generally produce several large litters of young per year and experience boom-and-bust population cycles.

Harvest mice—very small; tail and body not usually distinctly bicolored.

Deer mice—small; large eyes and ears; long, bicolored tail; white feet and belly.

Woodrats—medium-sized; large eyes and ears; long, sparsely haired to bushy tail; pale belly and feet.

Voles—small; indistinct, rounded head-body shape; small eyes and ears; stubby to medium tail.

Animal Sign: Collections of sticks, leaves, and objects stuffed in a crevice or beneath a boulder; small, conical lodge of mud and aquatic vegetation (muskrat); dark, hard, ricelike pellets (mice); larger, oval pellets (rats); scat/urine solidified in a dark, shiny mass at toilet site (woodrats).

Western Harvest Mouse
Reithrodontomys megalotis

Field ID: This mouse is buff to gray-brown with a wash of black or darker hair down the back, and a pale belly. The long, hair-covered tail is dark on top, white underneath.

Size: Head/body length 2⁴/₅–3 inches (71–76 mm), tail 2¹/₃–3¹/₅ inches (59–81 mm), weight ¹/₃–³/₅ oz (9–17 g).

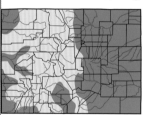

Habitat: Grasslands with dense, tall stands of grass, shrublands, weedy areas, riparian areas.

Distribution: Throughout eastern Colorado, the San Luis Valley, low elevation valleys in western Colorado.

Field Notes: On the Eastern Plains, western harvest mice inhabit moister riparian habitats and the edges of wetlands as well as grasslands with dense growth. They tend to live in drier habitats on the Western Slope. They also live at the edges of farm fields and in weedy pastures and roadsides. They feed on the seeds of grasses and forbs and frequently eat insects. Unlike many species of mice, they live in groups and huddle together in their dens in cold weather to keep warm. They don't hibernate, but in very cold weather they may go into brief periods of torpor. Their nests are balls of soft grass on the ground surface or in a shrub or even within an old woodpecker hole. Females are ready to breed at the age of three to four months and produce two to five litters, with an average of four young per litter, each season.

Legal Status: Nongame.

Plains Harvest Mouse
Reithrodontomys montanus

A photo of this species is not available. See the photo of the western harvest mouse on page 134 for a similar species.

Field ID: This small mouse is gray-brown with white undersides and hind feet and a long, haired tail with a narrow, black stripe along the top.

Size: Head/body length 2 ⅕–3 inches (56–76 mm), tail 2—2⅗ inches (51–66 mm), weight ⅕–⅓ oz (6–9 g).

Habitat: Dry grasslands.

Distribution: Throughout the Eastern Plains.

Field Notes: The plains harvest mouse is a small, slender mouse that is difficult to distinguish from the western harvest mouse and from deer mice. Harvest mice differ from deer mice by having grooves in their upper incisors, a feature only the most sharp-eyed naturalist is likely to notice. Plains harvest mice eat seeds as well as some berries, green vegetation, and insects, particularly grasshoppers. They are nocturnal. Their nest is a ball of soft grass and plant fibers set in a shrub or tree, inside a tree cavity, beneath a rock, or in or under an object. They are active year-round. Females produce multiple litters, each averaging four young, throughout the year. Young can breed at two months old. Though found throughout the Eastern Plains, plains harvest mice seem not to be abundant. Their populations may be impacted by heavy livestock grazing.

Legal Status: Nongame.

Brush Mouse
Peromyscus boylii

Field ID: This large mouse is gray-brown to buff-brown with tawny, orange-brown sides, pale undersides, and small ears. The well-haired tail is the same length as, or slightly longer than, the head and body. It is dark on top, pale underneath.

Size: Head/body length 3³/₅–4¹/₅ inches (91–107 mm), tail 3³/₅–4²/₅ inches (91–112 mm), weight ¾–1¼ oz (21–35 g).

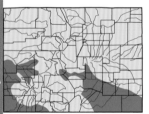

Habitat: Shrublands, especially oakbrush; riparian areas.

Distribution: Southeastern Colorado and southwestern Colorado west of the San Juan Mountains.

Field Notes: Brush mice are animals of arid, southwestern habitats. They live in shrublands of oakbrush and mountain mahogany and in pygmy forests of piñon pine and juniper. They are nocturnal, climbing into shrubs and trees to gather acorns, pine nuts, and juniper berries. They also feed on prickly pear cactus. In some areas, insects comprise up to half of their diet. The nests of most mice are a jumble of plant fibers with entrances on the sides or bottom, but brush mice build open-topped nests similar to bird nests. The nests are hidden under rocks or in crevices along cliffs and rock faces. Brush mice may build their nests within the enormous, jumbled nests of woodrats, even if the woodrat nest is occupied. The abundance of food affects how many litters a female has. Litters average three young.

Legal Status: Nongame.

Canyon Mouse
Peromyscus crinitus

Field ID: This beautiful, pale-colored mouse has long, silky fur and is buff-orange on the back, brighter buff on the sides, with white belly and feet. There may be an orangish spot on the breast. The long, haired tail is dark on top, white below, and has a slight tuft at the end.

Size: Head/body length 3–3$\frac{1}{3}$ inches (76–85 mm), tail 3$\frac{1}{2}$–4$\frac{1}{3}$ inches (89–110 mm), weight $\frac{2}{5}$–1 oz (11–28 g).

Habitat: Canyons, rock faces and rock falls, slickrock amid shrublands and piñon-juniper woodlands.

Distribution: Lower elevations of extreme western Colorado.

Field Notes: The canyon mouse is well named, living among the rugged canyons, cliffs, rock falls and broken country of western Colorado. It is agile and good at climbing the boulders and cliffs of its habitat. Well-adapted for life in this dry land, it feeds on seeds, berries, fungi, and insects and does not need free water. In summer and fall, insects are an important food, succeeded by seeds during the cold months. When food is scarce, the canyon mouse can gear down its metabolism and go into a daily torpor to conserve energy. Canyon mice are nocturnal, emerging to forage when they are less visible to predators. Females produce several litters a year of two to four young. The newborn young are heavier than those of other species and take longer to develop, being nursed for four weeks by the female, longer than many other mice.

Legal Status: Nongame.

White-footed Mouse
Peromyscus leucopus

Field ID: This mouse is gray, buff or reddish brown with a white belly and feet and a long, haired tail that is dark on top and pale beneath. The tail is shorter than the head and body.

Size: Head/body length $3^3/_5$–$4^1/_5$ inches (91–107 mm), tail $2^2/_5$–4 inches (61–102 mm), weight $1/_2$–$1^1/_{10}$ oz (14–31 g).

Habitat: Riparian areas, stream bottoms.

Distribution: Southeastern Colorado.

Field Notes: The white-footed mouse can be difficult to distinguish from the deer mouse, but its range in Colorado is restricted to the southeastern corner of the state while the somewhat smaller deer mouse is found state-wide. White-footed mice are good climbers and easily scale trees and shrubs. If discovered, they scramble quickly away up the trunk and along the branches. In spite of their willingness to climb, they build their nests on the ground from hair, feathers, shredded bark, plant down, leaves, and stems. White-footed mice are not selective in their food choices and insects are a major part of their diet, as well as seeds and other plant matter. In cold times or when food is scarce, they go into bouts of torpor lasting a few hours. They build their nests in old bird or squirrel nests, logs, stumps, objects, and buildings. They breed throughout the year, producing multiple litters averaging four young.

Legal Status: Nongame.

Deer Mouse
Peromyscus maniculatus

Field ID: This common native mouse can be reddish to gray-brown or buff, with white undersides and feet. It has a long, haired tail that is dark on top and white underneath. The tail is noticeably shorter than the head and body.

Size: Head/body length 2⅘–4 inches (71–102 mm), tail 2–5 inches (51–127 mm), weight ⅔–1¼ oz (19–35 g).

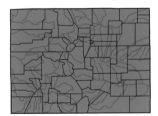

Habitat: Grasslands, shrublands, farm fields, vacant lots, urban lands, riparian areas, mountain forests, piñon-juniper woodlands, canyons, valley bottoms.

Distribution: Statewide.

Field Notes: Deer mice are found throughout most of North America, except for the Deep South, from the boreal forests of Canada down into central Mexico. They are by far the most common mammal in Colorado. Highly adaptable, they take advantage of a wide range of cover, from rock crevices and trees to leaf litter to human structures. They will eat whatever they can find, including seeds, nuts, plant matter, leaves, bark, insects, carrion, even bones. The species name, *maniculatus*, means "little hand," a nod to the mouse's dexterous front paws that can hold seeds and other objects. Given the tiniest access hole or crack, deer mice readily invade cabins, houses, farm buildings, sheds, and even vehicles. They breed year-round, depending on food availability and local conditions, producing several litters averaging eight young, which are able to breed at seven or eight weeks of age. Deer mice are active throughout winter. They move around among various nest sites along well-worn paths.

Legal Status: Nongame.

Northern Rock Mouse
Peromyscus nasutus

Field ID: This large, grayish brown mouse has whitish or silvery underparts, large ears, and a long, haired tail that is dark on top and white underneath and the same length as the head and body.

Size: Head/body length 3³/₅–4½ inches (91–114 mm), tail 3³/₅–4²/₅ in (91–112 mm), weight ⁵/₆–1 oz (24–28 g).

Habitat: Among shrublands in canyons, rock falls, rocky hillsides, and cliffs.

Distribution: The foothills of the Front Range and the Sangre de Cristo Mountains, east along the mesas of southeastern Colorado, and west into the San Luis Valley.

Field Notes: The rock mouse is well named, living its life closely tied to the rocky cliffs and canyons of the foothills east of the Continental Divide. It makes good use of this rugged landscape, finding abundant cover and nest sites under rock ledges and boulders, climbing, scampering, and running up, over, under, and around the rocky landscape. The juniper, piñon pine, oakbrush, and other shrubs of this habitat offer additional cover and abundant seeds, berries, and acorns for food. Rock mice breed in spring and summer, producing multiple litters averaging four young. The rock mouse can be distinguished from the more abundant deer mouse by its longer tail (about the same length as head and body), larger ears, and distinctly grayish fur.

Legal Status: Nongame.

Piñon Mouse
Peromyscus truei

Field ID: This large, buff to gray-brown mouse has very big ears, long, soft fur, white or pale gray feet and undersides, and a long, haired tail that is dark on top and white below and about the same length as the head and body.

Size: Head/body length $3^3/_5$–4 inches (91–102 mm), tail $3^2/_5$–$4^4/_5$ inches (86–122 mm), weight $^2/_3$–1 oz (19–28 g).

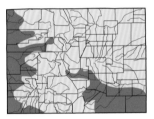

Habitat: Piñon-juniper woodlands.

Distribution: Southeastern Colorado south of the Arkansas River, west along the Arkansas to about Salida, western Colorado in mesa country and foothills and eastward along the Colorado River to Eagle County.

Field Notes: There is no confusing the piñon mouse with others of its genus because of its very large ears, which can be as much as an inch long. Closely tied to piñon pine/juniper habitat, piñon deer mice readily climb these low trees to reach food and shelter sites. Juniper berries and piñon nuts are their principal foods. Their nibbled leavings in the crotch of a tree or along a wide, level area on a juniper branch, along with a collection of mouse pellets, are sure signs of a favorite spot where a piñon mouse has stopped to munch a meal. Piñon mice also feed on flowers, fungi, and the occasional insect or carrion. Inhabiting arid habitats, they deal with drought by becoming torpid when water is limited. They are nocturnal and active year-round. While many *Peromyscus* species are aggressive and territorial, piñon mice are tolerant of other mice and don't mind crowded conditions. They breed spring through fall, producing several litters averaging four young.

Legal Status: Nongame.

Northern Grasshopper Mouse
Onychomys leucogaster

Field ID: This short-tailed mouse is gray to reddish brown with white underparts. It has short fur, a blocky body, and a short, haired, white-tipped tail.

Size: Head/body length 4–5 inches (102–127 mm), tail 1–2²⁄₅ inches (25–61 mm), weight 1–1²⁄₅ oz (28–40 g).

Habitat: Dry grasslands and shrublands.

Distribution: Eastern Colorado from the Front Range corridor to the Kansas state line, west along the Arkansas River, the San Luis Valley, far northwestern Colorado, along the Colorado River west of Grand Junction, western valleys along the Utah state line.

Field Notes: Many mice will take advantage of a protein meal of insects or carrion, but the grasshopper mouse actively hunts. Belying the stereotype of a timid mouse, the species has abandoned most plant foods, with the exception of seeds, in favor of a carnivorous diet of insects and any other small mice, shrews, birds, or reptiles it can capture and kill. This hunter lifestyle has led to behaviors more familiar in carnivores than rodents. Grasshopper mice have a single mate, both male and female care for the young, they mark territories, and they use a variety of sounds to communicate with each other. When the mated pair hunts, they keep in contact with birdlike chirps and twitters. They will rear on their hind legs and make a series of high-pitched squeaks, like tiny coyotes announcing their presence. Grasshopper mice breed March through September, producing three or four litters a year of one to six young. Females don't reach sexual maturity until they are six months old. They are active year-round, mostly at night.

Legal Status: Nongame.

Hispid Cotton Rat
Sigmodon hispidus

Field ID: This medium-sized native rat is gray-brown to blackish brown, with buff or yellowish hairs intermixed. It has pale undersides and a sparsely haired tail that is shorter than the head and body. The long, coarse fur nearly conceals the furred ears.

Size: Head/body length 5–8 inches (13–20 cm), tail 3–6 inches (76–152 mm), weight 4–7 oz (113–198 g).

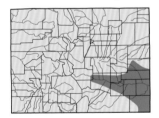

Habitat: Wetlands and grasslands with heavy cover.

Distribution: Southeastern Colorado and along the Arkansas River to Pueblo.

Field Notes: *Hispid*, meaning bristly, well describes the long, coarse, salt-and-pepper coat of the cotton rat. This native rat is an animal of midgrass and tallgrass prairie and weedy riparian habitat, where it finds good cover and food among the tall vegetation. It feeds on grasses and some forbs as well as insects, an occasional bird egg and even nestling mice. The cotton rat is mainly nocturnal or crepuscular, though it is sometimes active during the day. It creates runways along the ground surface that are often marked by cut grasses and vegetation. Cotton rats make a soft nest of grass set within a burrow or under a rock or log. They breed year-round and may produce up to 15 young in one litter, creating population booms in years of abundant resources. This species has been expanding its range from southeastern Colorado upstream along the Arkansas River and its tributaries.

Legal Status: Nongame.

White-throated Woodrat
Neotoma albigula

Field ID: This medium-sized woodrat has short, soft brown fur with a blackish wash, white throat, chest, and feet and large ears. The finely haired tail is dark on top, white beneath, and shorter than the head and body.

Size: Head/body length 7½–8½ inches (19–22 cm), tail 5½–7⅓ inches (140–186 mm), weight 4⅘–10 oz (136–284 g).

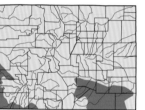

Habitat: Canyons, shrublands, piñon-juniper woodlands, grasslands, riparian areas, rocky slopes.

Distribution: Southeastern Colorado and far southwestern Colorado.

Field Notes: This woodrat inhabits dry, western habitats of rocky shrublands, grasslands, and canyons. Its life history is associated with cholla, prickly-pear, and other cacti. Cactus is its food of choice, though it also eats yucca leaves, juniper needles, and parts of piñon pine, sagebrush, and other shrubs, grasses, and desert vegetation. The spiny cover of cholla or other shrubs protects the den, which is built of cactus joints. Specific adaptations to life in the desert allow the white-throated woodrat to survive in an extreme habitat. It is able to metabolize the oxalic acid in cactus, which is toxic to most animals, and doesn't need to drink free water because of the high moisture content of the succulent plants (cactus) it eats. White-throated woodrats breed January through July, producing multiple litters averaging two or three young. They may have more than one litter in the nest at one time. Like other woodrats, it is nocturnal and active year-round. Also true to its pack rat reputation, it builds a big, messy nest of sticks, rocks, and various accumulated items, which may include shiny objects, toys, and trash in areas near human activity.

Legal Status: Nongame.

Bushy-tailed Woodrat
Neotoma cinerea

Field ID: This large woodrat is buff to gray, with paler sides, white undersides, and white feet. The ears are large and the long, bushy tail is almost squirrel-like. The soles of the hind feet are furred.

Size: Head/body length 7–9⅔ inches (18–25 cm), tail 5–7⅔ inches (13–19 cm), weight 7½–20¼ oz (213–574 g).

Habitat: Talus slopes at and above timberline, rocky areas within coniferous mountain forest, shrublands, and piñon-juniper woodlands, canyons, around old mines.

Distribution: From eastern foothills west to the Utah state line.

Field Notes: This woodrat is larger and has bigger ears than the other species of woodrats. It is easily recognized by its bushy tail and hairy hind feet. Better known by its nickname, "pack rat," this species entered western folklore during the heyday of western gold and silver mining because of its habit of stealing shiny trinkets and trash and adding them to its nest. The woodrat's "house" is a collection of sticks, leaves, bones, and appropriated items piled into an enormous structure beneath a boulder, rock overhang, fallen tree, or other solid protection. The actual nest is usually made of soft plant fibers and tucked within this collection. An accumulation of droppings and a shiny black varnish that forms over years from urine also mark the woodrat's habitation. Woodrats eat the leaves, needles, and fruits of juniper, piñon pine, chokecherry, mountain mahogany, and other mountain shrubs and forbs. The bushy-tailed woodrat is nocturnal and adept at climbing rock faces and the timbered walls of old cabins. It breeds April through August, producing two litters a year, averaging three to four young.

Legal Status: Nongame.

Eastern Woodrat
Neotoma floridana

Field ID: This large woodrat is grayish brown with a white or pale gray throat and underparts. The sparsely furred tail isn't scaly, is shorter than the head and body, and is dark on top and white or pale gray beneath.

Size: Head/body length 8–9 inches (20–23 cm), tail 6–8 inches (15–20 cm), weight 7–13½ oz (198–383 g).

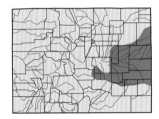

Habitat: Riparian areas, shrublands.

Distribution: Central Eastern Plains and west along the Arkansas River to Pueblo.

Field Notes: Colorado is at the far western edge of the eastern woodrat's range. While the more western woodrat species are adapted to deserts, canyons, or mountains, the eastern woodrat is found in and around deciduous woodlands, making its home in cottonwood and box-elder groves along waterways of the Eastern Plains and in areas with large shrubs of rabbitbrush, chokecherry, snowberry, and yucca. It eats the leaves, bark, tender twigs, berries, and fruit of yucca, snowberry, currant, wild rose, rabbitbrush, chokecherry, and other shrubs, trees, and forbs. Like other woodrats, it builds a large, jumbled house of sticks, bones, and any other collectibles. It is a solitary, nocturnal animal, active year-round. As it moves to and from its house on nightly foraging trips, the animal may shove aside accumulated debris, wearing a path around the house. It caches food and does not need to drink free water. It breeds in spring and summer, the female bearing two or three litters averaging three young each.

Legal Status: Nongame.

Desert Woodrat
Neotoma lepida

Field ID: This small woodrat is yellowish to buff-gray, with white or buff underparts. The tail is distinctly dark on top and white below and covered with short hairs.

Size: Head/body length 6–7 inches (15–18 cm), tail 4⅓–6½ inches (110–165 mm), weight 3⅓–6 oz (94–170 g).

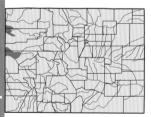

Habitat: Canyons, desert shrublands.

Distribution: Far western Colorado along the Colorado and White rivers.

Field Notes: Colorado lies at the far eastern edge of the desert woodrat's range. This is an animal of western canyons and shrublands of the Great Basin. Making use of this broken, rocky country, desert woodrats build the typical pack rat house—a chaotic collection of sticks, bones, and other found items—in rock crevices and under boulders. This woodrat feeds on the seeds, acorns, and fruits of desert shrubs and forbs and doesn't rely as heavily on cactus as other woodrats. Like other woodrats, it derives moisture from its food and doesn't need to drink free water. Nocturnal and solitary, desert woodrats have been reported to vibrate their tails against dry vegetation, mimicking the rattle of a rattlesnake, to frighten away predators. Desert woodrats breed in spring and summer. They have two litters of about two young, which are able to breed at three months old.

Legal Status: Nongame.

Mexican Woodrat
Neotoma mexicana

Field ID: This medium-sized woodrat is pale to dark gray with a wash of tawny or black. The underparts are grayish white. The haired tail is about the length of the head and body and is black on top and white below.

Size: Head/body length 6½–7¾ inches (17–20 cm), tail 6–6½ inches (15–17 cm), weight 5¼–9 oz (149–255 g).

Habitat: Canyons, pine forests, piñon-juniper woodlands, shrublands, rocky slopes, cliffs.

Distribution: The foothills of the Front Range, west along the Arkansas River to Salida, southeastern Colorado, and southwestern Colorado.

Field Notes: Smaller size, moderately furred tail, and a lack of hair on the feet help distinguish this animal from the bushy-tailed woodrat. It inhabits foothills, broken rocky areas, and canyons rather than the high mountains preferred by the bushy-tailed woodrat. The Mexican woodrat gathers its pack rat collection of sticks and "stuff" into a rock crevice or beneath a boulder rather than in an aboveground house like some other woodrats. It may also move into old cabins or under porches and sheds. Abundant deposits of brown, lozenge-shaped pellets mark the site of its activity and its regular toilet spots. Mexican woodrats eat a great variety of plant material, including the leaves of oakbrush, sagebrush, chokecherry, and other shrubs as well as pine nuts, juniper berries, and acorns. They cut and store green foliage. Branches with dried green leaves often mark their dens. They are less efficient than other woodrats at extracting moisture metabolically and will drink free water. Active year-round and nocturnal, they produce two litters a year averaging three young.

Legal Status: Nongame.

Southern Plains Woodrat
Neotoma micropus

Field ID: This medium-sized, steel gray woodrat has a gray belly and white throat, breast, and feet. The haired tail is slightly shorter than the head and body and black above, gray below.

Size: Head/body length 7½–8½ inches (19–22 cm), tail 5½–6½ inches (14–17 cm), weight 7–11 oz (198–312 g).

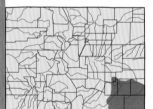

Habitat: Grasslands, rocky outcrops, and bluffs of shrublands.

Distribution: Far southeastern Colorado, mostly south of the Arkansas River.

Field Notes: Like its cousins, this woodrat builds the collection of sticks and debris that define a pack rat nest, set on a rock outcrop or at the base of a shrub. Within this "house" are chambers for sleeping, rearing of young, and food storage. The woodrat often uses parts of cholla and prickly pear cactus in building the house. It is a solitary, nocturnal animal, emerging at twilight to forage for food. Its main food is cholla, a dominant shrub in the lower Arkansas River basin of Colorado. Southern plains woodrats also eat grass, yucca, prickly pear, juniper, wild currant, and other shrubs and forbs. They breed in spring, producing one or more litters a year of two to three young. They are reported to communicate with other woodrats by drumming with their hind feet. Like all the woodrats, they rely on their large ears and sensitive whiskers to gather information in the dark.

Legal Status: Nongame.

Southern Red-backed Vole
Myodes gapperi

Field ID: This dark gray to gray-brown vole has a distinctive panel of red from the forehead to the tail, pale gray or whitish undersides and a short tail that is dark on top, light underneath. The body is plump, the nose fairly pointed, and the small ears mostly hidden in the fur.

Size: Head/body length 3²/₃–4²/₃ inches (93–119 mm), tail 1–2 inches (25–51 mm), weight ½–1²/₅ oz (14–40 g).

Habitat: Mountain coniferous forests, aspen forests, mountain meadows, talus slopes, willow riparian areas.

Distribution: Central Colorado throughout the mountainous portion of the state.

Field Notes: This subalpine-forest vole makes good use of the enormous middens of its neighbor, the chickaree, using these collections of cones and scales for food and cover. It will eat whatever is available, including seeds, berries, fruits, leaves, fungi, and bark. It remains active through winter, sheltered in sub-surface runways and dens. This species breeds from late winter through fall, producing two to three litters averaging six young. The nest is a round ball of soft grass or moss. The red-backed vole is a good climber, a convenient skill for a forest-dweller. During the warm months it is crepuscular and nocturnal, but in winter, when it can move about beneath the snow without being seen by predators, it becomes active during the day.

Legal Status: Nongame.

151

Western Heather Vole
Phenacomys intermedius

Field ID: This small vole is gray to brown, with long, soft fur, silvery underparts, and white feet. The stubby tail is dark on top and pale underneath with white hairs speckling the dark upper side.

Size: Head/body length 3½–4⅗ inches (89–117 mm), tail 1–1⅔ inches (25–42 mm), weight 1–1⅖ oz (28–40 g).

Habitat: Coniferous mountain forests, aspen forests, mountain meadows, alpine tundra.

Distribution: Through the central mountains up to about 12,000 feet.

Field Notes: The heather vole is also called the mountain vole. It is an animal of northern North America, distributed from the Rocky Mountains northward and across forested regions of Canada to the Arctic Circle. In the Rockies it inhabits subalpine forests and alpine tundra, usually near water. Heather voles are mainly nocturnal or crepuscular, foraging for seeds, berries, tree bark, fungi, and green vegetation. In summer, they inhabit burrows but in winter build well-insulated nests on the ground surface, beneath the snow. They don't hibernate but store food in their burrows for winter. Though usually solitary animals, family groups have been found huddled together in communal nests in winter. Females produce multiple litters of five or six young.

Legal Status: Nongame.

Long-tailed Vole
Microtus longicaudus

Field ID: This small vole is reddish brown to gray-brown with black hairs intermixed, paler undersides, and whitish feet. The long tail is close to half the length of the head and body and distinctly dark on top and pale beneath.

Size: Head/body length 4½–5⅓ inches (114–135 mm), tail 2–3½ inches (51–89 mm), weight 1⅓–2 oz (38–57 g).

Habitat: Aspen forests, wet meadows, mountain grasslands, alpine tundra, subalpine forests, sagebrush shrublands.

Distribution: Across the western half of the state from the eastern foothills westward.

Field Notes: Long-tailed voles inhabit a variety of mountain habitats, from forests, willow thickets, and subalpine meadows, to semiarid shrublands of western Colorado. They feed on grasses, green vegetation, fruits, and seeds, switching in winter to the bark, twigs, and buds of trees and shrubs. While most voles build elaborate runways through vegetation and beneath the snow, the long-tailed vole is not dependent on runways, and the ones it builds aren't well defined. This allows the animals to make use of drier habitats with less vegetation and cover. Long-tailed voles breed prolifically, producing several litters a year, each with up to seven young. While the tails of most Colorado voles are short or even stubby, this one's long tail helps distinguish it.

Legal Status: Nongame.

Mexican Vole
Microtus mexicanus

Field ID: This short-tailed vole is dark brown with buff underparts. The tail is less than one-third the length of the head and body. The small round ears are covered with fur.

Size: Head/body length 4–4⅗ inches (102–117 mm), tail 1–1⅖ inches (25–36 mm), weight 1–1½ oz (28–43 g).

Habitat: Grassy areas of ponderosa and piñon-juniper woodlands and mountain shrublands.

Distribution: Western Las Animas County and the Four Corners region.

Field Notes: Not a lot is known about this species in Colorado. Because of the small number of records, the Mexican vole was thought to have a limited range in the state but is now suspected to inhabit much of southeastern Colorado. While many voles are nocturnal, this species has two activity periods, one during midday and the other in early evening. Mexican voles build globe-shaped nests of dry grasses set under logs, dense vegetation, or in an excavated burrow. They breed through the summer but have small litters for voles, averaging two young. Mexican voles live in colonies, and their many burrows wind through tall grasses, under shrubs, and around rocks. They feed on green grasses and forbs in summer, switching to roots, bulbs, tender bark, and softer parts of perennial plants during the winter. They don't hibernate, and they don't store food for winter, instead foraging year-round along the same paths through their extensive burrow systems.

Legal Status: Nongame.

What Is a Vole?

They're not moles. They're not mice. They're voles, a group of small rodents that few people ever actually see. That doesn't mean they aren't important. Voles are found in nearly every habitat in Colorado. These small animals live where the soil meets the vegetation, scurrying through runways chewed through the grass. The word *vole*, in fact, comes from the Old Norse word *vollr*, meaning "field." Voles are eaten by just about every kind of predator, so they're a crucial prey base for Colorado's ecosystems. Nicknamed meadow mice, they don't look much like typical mice. They have rounded bodies with small ears often hidden within their thick fur. Their eyes are small, and their tails are usually short to stubby to almost invisible. The reduction of external body parts is probably an adaptation for their life at the ground surface, where protruding features would just get in the way.

In spring, when the snow melts, the secret life of the vole is revealed. "Half-pipe" runways chewed through the grass and snaking extensively hither and yon show how the vole spent its winter foraging under the snow. But as the days warm in spring, the concealing roof of these winter tunnels melts away with the thaw.

Montane Vole
Microtus montanus

Field ID: This large vole is gray-brown to black with silvery underparts. The tail is less than half the length of the body and head, and the small ears are hidden in the fur.

Size: Head/body length 4–5½ inches (102–140 mm), tail 1⅕–2⅗ inches (30–66 mm), weight 1–3 oz (28–85 g).

Habitat: Wet meadows, wet grasslands, aspen stands, drier grasslands.

Distribution: Throughout the mountains of the western half of Colorado from about 6,000 feet to above timberline.

Field Notes: Like little lawnmowers, montane voles clip endless runways through the grass of mountain meadows, leaving crisscrossing trails on the ground surface beneath the vegetation. Spread apart the grassy thatch almost anywhere in these meadows and you will likely find the trail of a vole. Voles are mainly nocturnal and active throughout the year, moving about within their runways under protection of cover and/or snow. They have a specialized diet, eating mostly the leaves of wildflowers and other forbs. Though close in appearance to other voles, the montane vole is larger and stockier than the western heather vole, has a shorter, less bicolored tail than the long-tailed vole, and occupies different habitats than the prairie or meadow vole. Montane voles breed in the warmer months. Their populations cycle through highs and lows and may increase, peak, and crash within three to four years. Because of the short growing season of their mountain habitat, these voles have large litters of up to ten young, but fewer litters per year than other species. Young born early in the summer breed later that year.

Legal Status: Nongame.

Prairie Vole
Microtus ochrogaster

Field ID: This medium-sized vole is gray to reddish brown with pale buff or gray underparts. The short tail is dark on top, pale beneath.

Size: Head/body length 3½–5 inches (89–127 mm), tail 1⅕–1⅗ inches (30–41 mm), weight 1–1½ oz (28–42 g).

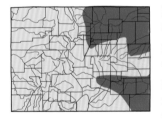

Habitat: Grasslands, edges of farm fields, sagebrush shrublands.

Distribution: Northeastern and southeastern Colorado.

Field Notes: The prairie vole is aptly named, being found throughout the central prairies of the United States and Canada and eastward into Ohio. In Colorado, it is an animal of the Eastern Plains. While many voles are intolerant of others of their kind, prairie voles are social. They live in colonies of extended families of multiple generations as well as unrelated animals. Adult males and females form monogamous pairs and together rear the young and keep the burrows and runways clipped. They breed in the warm months, producing multiple litters averaging three to four young. As grassland dwellers, they eat grass but cannot survive on a grass-only diet. Other foods include roots, stems, leaves, bark, and insects. Where the grass is thick, prairie voles construct elaborate runways, but in drier habitats they build shallow burrows. Their nests, made of grass and shredded bark and containing large caches of stored roots and seeds, are secreted within the burrows. Prairie voles are active year-round and at any time of day or night, but in summer are more crepuscular and nocturnal to avoid predators.

Legal Status: Nongame.

Meadow Vole
Microtus pennsylvanicus

Field ID: This large vole is gray to dark brown or chestnut with silvery underparts. The long tail is sparsely haired and darker on top, pale underneath. The round ears are hidden in the fur.

Size: Head/body length 3½–5 inches (89–127 mm), tail 1²/₅–2³/₅ inches (35–66 mm), weight 1–2½ oz (28–71 g).

Habitat: Wet meadows, wetlands, riparian areas, marshes.

Distribution: Through much of the mountains and valleys of central Colorado, along the South Platte River from its headwaters to northeastern Colorado, along the Republican River, and in the San Luis Valley.

Field Notes: The meadow vole is the most widespread, and probably the most-studied, vole in North America. It is a creature of moist places—wet meadows, marshes, riparian areas—and it eats the grasses, sedges, rushes, and forbs found there. It is also a good swimmer. Where it shares range with other kinds of vole (montane, prairie, long-tailed), the meadow vole tends to inhabit moister places, the others occupying drier slopes or uplands. The meadow vole is active year-round and day or night, though its diurnal activity increases in winter with the cover of snow. Meadow voles create runways clipped through the vegetation. The species breeds year-round, and females produce multiple litters of up to 11 young, which grow quickly and can breed by six weeks of age. Populations cycle every few years. Numbers of animals in a hectare (about 2½ acres) of habitat may reach the hundreds, then drop again to only a few individuals over the same area.

Legal Status: Nongame.

Sagebrush Vole
Lemmiscus curtatus

Field ID: This small vole has thick, soft fur and a short, furry, stubby tail. It is paler than other voles, from gray to buff, with paler sides. The back part of the soles of the feet are furry.

Size: Head/body length 3⅘–4½ inches (97–114 mm), tail ⅗–1⅛ inches (15–29 mm), weight ⅘–1⅓ oz (23–38 g).

Habitat: Sagebrush shrublands.

Distribution: Northwestern Colorado, and North Park.

Field Notes: This vole is a denizen of the big-sagebrush shrublands of western Colorado. It feeds mainly on grasses but also the soft parts of sagebrush, rabbitbrush, winterfat, and other shrubs and forbs such as lupine and composites. It is active year-round, mainly late in the day and evening. It constructs runways that, because of the sparse vegetation of its habitat, are not as elaborate as those of other voles. Burrows are usually built beneath the cover of big sagebrush and contain nests lined with grass and leaves. Sagebrush voles are social and may live in large colonies. Though a mated pair stays together after the young are born, the male doesn't share in the care of the young. Like other voles, sagebrush voles are prolific and may have two to three litters a year, averaging five or six young.

Legal Status: Nongame.

Muskrat
Ondatra zibethicus

Field ID: This large rodent has a stocky body, round head and long, scaly tail that is flattened side-to-side. Its fur is red-brown to very dark brown. It has short legs, a rounded posture, dexterous front paws, and partly webbed hind paws.

Size: Head/body length 10–14 inches (25–36 cm), tail 8–11 inches (20–28 cm), weight 2–4 lbs (907–1,814 g).

Habitat: Marshes, wetlands, ponds, lakes, slow-moving streams, farm ponds.

Distribution: Statewide.

Field Notes: Though closely related to voles, muskrats are often mistaken for small beavers (and sometimes mink), with whom they share a semiaquatic lifestyle. A close look reveals the muskrat's long, sticklike tail that is flattened on the sides, allowing it to act as a rudder, unlike the larger beaver's wide, flat tail. The muskrat is well adapted to life in the water. Its fur—a dense, soft undercoat with a topcoat of long, coarse guard hairs— is almost waterproof. The front feet are dexterous, like a raccoon's, and the large hind feet are partially webbed to aid in paddling. The lips close behind the teeth and the nostrils can pinch shut to keep out water. Muskrats are abundant and widespread. They adapt well to life around people, building two-to-three-foot-high dome-shaped houses of rushes, sedges, and cattails, which look like mini beaver lodges, along ponds and streams. They also construct burrows in the bank. Muskrats are often seen as a low, round shape creating a vee in the water as they paddle across a quiet pond. They feed on marsh vegetation and occasionally on fish, crayfish, snails, and carrion. They are active year-round, groups of them huddling together in the nests for warmth in winter. Muskrats breed in the spring, producing two litters of four to eight young. Curiously, the young are born with round tails, which flatten on the sides by the time they are two months old. Muskrat populations cycle up and down over five- to ten-year periods.

Legal Status: Furbearer.

Good Rat, Bad Rat: Woodrat, Lab Rat

The term *rat* carries all kinds of negative connotations, conjuring images of nasty, disease-carrying vermin. These "Old World" rats, of the genus *Rattus*, are not native to North America but were introduced by humans. They also have been bred as laboratory rats.

Colorado's native rat species are very different from those unsavory, naked-tailed sewer or city rats. They are essential components of Colorado's wild ecosystems.

Kangaroo rats are more like big mice than the usual image of rats, with large eyes, long tails, and long, powerful legs. Despite their name, muskrats could be thought of more as scaled-down beavers, or scaled-up meadow mice. They live along streams and ponds and feed on aquatic vegetation. Woodrats (often called pack rats) are the same size and shape of Old World rats (genus *Rattus*) and have long tails, but can be distinguished in several ways:

- The throats and bellies of woodrats are white or very pale. Those of Old World rats are gray.
- The woodrat's tail is covered with hairs and the scales of the skin are not easy to see. Tails of Old World rats are clearly naked and scaled.
- Woodrats live in wild habitats. Old World rats rarely live far from people, their buildings and their trash.

Family Muridae—Old World Mice and Rats

The house mouse and Norway rat, which are not native to North America and were introduced by humans, are the prototype rodents of folklore and the public imagination. They live close to humans and often infest homes, buildings, and structures.

They have long, naked tails with the scales of the skin visible. The fur is short and the coloring is nearly uniform, not dramatically counter-shaded (dark on top, light beneath) as with native mice and rats.

Animal Sign: Hard, dark, ricelike droppings (rat droppings significantly larger); urine spots; musty odor; chewing and damage of foodstuffs; clothing and soft objects chewed up for nest material.

House Mouse
Mus musculus

Field ID: This small, slender, pointed-nosed mouse is pale to dark gray or brown, with a long, naked, scaled tail and large naked ears, small black eyes, and prominent whiskers. The coat is thin and silky. The musky odor is unmistakable.

Size: Head/body length 3–3½ inches (76–89 mm), tail 2½–4 inches (64–102 mm), weight ⅖–⅘ oz (11–23 g).

Habitat: Human structures, including houses, barns, farm buildings, silos, and warehouses, as well as old fields, ditches, and farmland (but mostly near buildings).

Distribution: Statewide.

Field Notes: Like the Norway rat, the house mouse is a nonnative species brought to North America in the baggage of European immigrants. Highly adaptable, it quickly spread continent-wide, thriving in and around humans. House mice make use of whatever food they can find, including grains, crops, vegetation, insects, carrion, and fruit. They are nocturnal and secretive, constructing nests of soft material in nooks and crannies. They can be very destructive, chewing into stored food, pillows, clothing, and household items for food and nesting materials. Their urine and abundant droppings are evidence of their infestation, as well as a characteristic musky odor, produced for scent marking by anal glands. House mice can be distinguished from native mice by their slender bodies, pointed noses, and naked ears and tail. While many native mice are dark on their backs and pale to white underneath, house mice are more uniformly colored. House mice breed continuously, producing five to seven litters of up to 13 young. The young can breed by six weeks old.

164 **Legal Status:** Nonnative, no protections.

Norway Rat
Rattus norvegicus

Field ID: This large, sturdy-bodied rat ranges in color from pale brown or gray to very dark. It has a short, coarse coat and prominent, nearly naked ears. The long tail is naked or very sparsely haired with evident scales.

Size: Head/body length 7–10 inches (18–25 cm), tail 5–8 inches (13–20 cm), weight 7–10 oz (198–284 g).

Habitat: Sewers, landfills, weedy fields, old or vacant buildings, feedlots, usually in or near cities and towns.

Distribution: In towns and cities of eastern Colorado, especially along the Front Range urban corridor.

Field Notes: The Norway rat is the notorious sewer rat reviled as a verminous pest. Norway rats carry many diseases and also contaminate and destroy food stores, crops, and property. They are not native to North America and arrived in the holds of ships from Europe. Once on this continent, these adaptable rodents spread quickly. In Colorado, they are found mostly in urban areas and around feedlots but are rarely as abundant as in older, more humid, and more densely populated cities of the United States. Norway rats are the ultimate opportunists and will eat almost anything, including garbage, carrion, and any plant material, as well as birds, small animals, insects, bird eggs, and anything they can catch and kill. They climb and swim very well, which allows them to make use of a variety of cover and food sources. They typically get in or under buildings, into sewers, ditches, trash and garbage dumps, and various structures. They live in colonies of a dozen or more animals, the females producing up to five litters a year averaging 7 to 11 young, which is why populations can boom where there are sufficient resources.

Legal Status: Nonnative, no protections.

Family Dipodidae—Jumping Mice and Jerboas

Jumping mice have extremely long tails that can be twice the length of the body. They have very long legs and hind feet, adaptations for jumping. The family name Dipodidae means "two-footed," a nod to their two-legged, bounding mode of locomotion.

Jumping mice live in areas with lush vegetation—meadows, creeksides, grassy slopes—but will move onto drier uplands to forage.

Animal Sign: Globelike nest of grass at the end of a burrow; piles of grass stems at feeding sites; tracks in streamside mud with elongated hind feet, four-toed front feet often between hind feet.

Meadow Jumping Mouse
Zapus hudsonius

Field ID: This small, long-tailed, pointed-nose mouse is yellowish brown on the back with yellow sides and white underparts. The hind legs and feet are quite long, and the sparsely haired tail is thin and twice the length of the body.

Size: Head/body length 3–3⅓ inches (76–85 mm), tail 4–6 inches (102–152 mm), weight ½–⅘ oz (14–23 g).

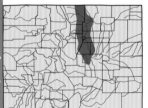

Habitat: Streamsides and adjacent uplands and riparian areas.

Distribution: Along the Front Range corridor, from Wyoming to El Paso County, to the edge of the Eastern Plains.

Field Notes: Though they superficially resemble house mice, jumping mice are quite different. Like mini-kangaroos, they have very long hind legs and big hind feet. They can leap 18 inches high and change direction in midair. The 6-inch tail attached to the 3-inch body provides a counterweight for these tremendous leaps. This nocturnal mouse shelters from predators during the day in a grassy nest hidden under streamside vegetation. At night, the mouse forages along the creek and climbs up onto the nearby upland searching for food. It eats grass and plant seeds, berries, spiders, caterpillars, even fungi. Jumping mice breed throughout the warm months, producing two litters averaging four to five young. Active from mid-May to mid-October, they hibernate in winter in underground burrows. The subspecies of meadow jumping mouse that inhabits Colorado is Preble's meadow jumping mouse (*Zapus hudsonius preblei*). The riparian habitat along streams of the Front Range corridor that it depends on has been greatly altered or destroyed by human development, and Preble's mouse is listed as threatened under the Federal Endangered Species Act. A second subspecies (*Zapus hudsonius luteus*) has been reported along the New Mexico border east of Trinidad and Raton Pass.

Legal Status: State and federal threatened species.

Western Jumping Mouse
Zapus princeps

Field ID: This medium-sized mouse has a very long, thin tail and pointed nose. The back is blackish mixed with yellow. It is yellowish or buff on the sides, with a white belly. The hind legs and feet are long and the sparsely haired tail is nearly twice the length of the body.

Size: Head/body length 3½–4 inches (89–102 mm), tail 5–6 inches (13–15 cm), weight ⅔–1⅓ oz (19–38 g).

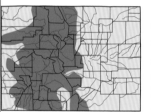

Habitat: Foothills and mountain riparian areas between 6,000 and 11,500 feet.

Distribution: From eastern foothills westward across the mountains and higher elevations of the state.

Field Notes: The very long tail and long hind feet of the jumping mouse distinguish it from other small-bodied mice. The western jumping mouse is larger than its cousin, the meadow jumping mouse. It is a nocturnal animal but is occasionally active during the day. Fly fishers sometimes see jumping mice foraging in the vegetation along mountain streams, its favored habitat. The mice sometimes surprise those same anglers by swimming, occasionally chasing the fishing lines. Western jumping mice must make the most of summer in the high country. They hibernate in winter, in deep burrows below the frost line. Those at high elevations enter their hibernacula by September, not to emerge until June. Perhaps because of the limited time spent active, they are long-lived for mice, living four or more years as compared to less than one year for many mice species. They breed in summer, producing one litter averaging five young.

Legal Status: Nongame.

Family Erethizontidae—New World Porcupines

This family is represented by a single species in North America (north of Mexico)—the curious and fascinating porcupine. Porcupines are slow-moving and near-sighted and would be easy prey, except for the outstanding defensive adaptation they are famous for. The unwitting predator (including the family dog) who makes a grab for a porcupine learns a painful lesson. The porcupine's outer guard hairs are modified as sharp, barbed quills. Porcupines walk on the soles of their feet, like raccoons and bears, and the soles have scaly surfaces that help them climb. They are stocky-bodied and have long claws for climbing.

Animal Sign: Large, bare oval or irregular patches without bark on tree trunks and limbs, with neat, chewed edges showing teeth marks; pigeon-toed tracks with claw marks showing ahead of the foot and a pebbly pattern on the sole, usually showing a tail drag; shed quills.

North American Porcupine
Erethizon dorsatum

Field ID: This large rodent is yellowish brown to blackish and stocky, with a short, thick tail and short legs. The body is covered with sharp, black-tipped white or yellowish quills and coarse guard hairs that give a white or yellow cast to the coat.

Size: Head/body length 18–22 inches (46–56 cm), tail 7–9 inches (18–23 cm), weight 10–28 lbs (5–13 kg).

Habitat: Foothills and mountain coniferous forests, riparian areas, aspen groves.

Distribution: Statewide, though most common in forested central and western Colorado.

Field Notes: The porcupine's quill coat makes it one of the most distinctive and recognizable animals in North America. The quills are actually modified hairs. Contrary to folklore, porcupines cannot shoot their quills, but they will erect their quills when threatened and swing their heavy tails to swat at a threatening animal. The loosely attached quills embed in the skin of any would-be predator curious or naïve enough to touch the "quill pig." Projecting barbs make the quills difficult to remove. Once they have learned the painful lesson of tangling with a porcupine, most predators will give it a wide berth. Slow-moving and near-sighted, porcupines spend their nights feeding on the inner bark, buds, leaves, or needles of trees and shrubs, especially pines and Douglas-fir, and their days sleeping in the branches of these same trees. Large, oval patches chewed into tree bark, usually high up on the trunk, are evidence of porcupine activity. They have excellent senses of hearing and smell and will rear on their hind legs to catch a scent. Porcupines are active year-round. Weighing up to 40 pounds, porcupines are the second-largest rodents in North America, after beavers. They are solitary animals, breeding in fall or early winter but not bearing their single (or rarely twin) young until spring. The gestation period is an astonishing seven months, extremely long for a rodent. Young are born with soft quills, which harden quickly after birth.

Legal Status: Small game.

ORDER
CARNIVORA

Carnivores: Meat-eaters

The carnivores are a broad group of species, including the cat, dog, bear, raccoon, skunk, and weasel families. In spite of their name, many are not strictly meat-eaters. Coyotes and foxes are omnivores, eating whatever they find. Vegetation comprises a large percentage of the diet of bears. Carrion is also an important food source for many carnivores. Cats, on the other hand, are strict meat-eaters and eat almost exclusively live prey.

Carnivores have five toes on the front foot. Cats and dogs have only four toes on the hind foot, and the fifth toe of the front foot is a dewclaw located on the leg above the paw. Thus cat and dog tracks show only four toes on all feet. Bears, weasels, skunks, and raccoons have five toes on the back foot and five toes usually show in their tracks. With the exception of bears, carnivores do not hibernate (see Sleeping through the Season sidebar on page 100), though some enter periods of reduced activity during winter.

All carnivores have canine teeth. Carnivores occupy the upper level of the food web, and therefore they have smaller populations than the prey animals they feed on.

Family Canidae—Dogs

The members of this family are best recognized by their resemblance to domestic dogs. Coyotes and wolves are in the same genus as domestic dogs, *Canis*. Dogs are domesticated wolves, though humans have altered their appearance through selective breeding (artificial selection) into a huge variety of shapes and sizes.

Canids have pointed muzzles; large, upright ears; large eyes; long legs; and long, bushy tails. They walk on their toes (digitigrade) and show four toes in each track, though the front paws have a fifth toe, the dewclaw.

Members of the dog family are intelligent, mobile, adaptable, and resourceful. Animal flesh is a major part of their diet, but most also eat berries, nuts, and whatever food is available. They are well adapted to various climates, from high mountains to deserts. Coyotes and red foxes adapt especially well to life around people, changing their habits to make use of human-created habitat rather than being displaced.

Some canids, like wolves, are highly social. All provide a high degree of care of the young and communicate with various vocalizations, facial expressions, body postures, and scents. They produce one litter of pups a year, and their dens are usually dug beneath a bank, boulder, or other cover.

Animal Sign: Four-toed paw track longer than it is wide, sometimes showing a claw mark in front of each toe pad; front track larger than hind; doglike scat; dens dug in a bank or under concealing object; musky, skunklike odor (foxes); yipping and howling.

Coyote
Canis latrans

Field ID: This medium-sized member of the dog family is grizzled gray with golden, red, and black intermixed. Its underparts are pale to white. It is slender-bodied, with long, slim, reddish brown legs, pointed muzzle, upright ears and a long, bushy tail.

Size: Head/body length 32–37 inches (81–94 cm), tail 11–16 inches (28–41 cm), weight 20–50 lbs (9–23 kg).

Habitat: Grasslands, agricultural areas, open mountain forests, mountain parks, piñon-juniper woodlands, shrublands, urban/suburban open space and parks.

Distribution: Statewide.

Field Notes: Coyotes are amazingly adaptable and opportunistic, eating almost anything and living almost anywhere. They hunt small mammals and eat nuts, berries, fruit, insects, and carrion. They may occasionally kill live prey as large as deer, especially old, sick, or winter-weakened animals. They rarely hunt in packs, but two coyotes may work together. Using its large ears and acute hearing, the hunting coyote listens for rodents in the grass, then rears on its hind legs and pounces, pinning its prey with its front paws. Despite intense eradication efforts, the species has expanded its range in the last century, moving to exploit habitat and ecological roles left vacant by human extirpation of the gray wolf. Coyotes are now found throughout North America. Studies have shown that intense control efforts that reduce local populations result in increased litter size among coyotes. Coyotes adapt well to life around humans, living not only in wild areas and rural habitats, but also moving into suburbs and even some urban areas. Pairs mate January through March, then excavate a den with good cover. The female gives birth to five or six pups between March and May. The male feeds the nursing female, then both hunt for the young, often helped by adult young from earlier litters. By late summer, the young are old enough to accompany the adults on hunts. The extended family gathers for a group sing to reaffirm their group bonds before heading out to hunt. The lonely howl of the coyote, nicknamed "song dog," is an iconic symbol of the American West. In addition to howling, coyotes communicate with yips, short barks, tail and body posture, scent, and facial expression. In fall, the family disperses. Coyotes may live in groups or singly, depending on available resources and conditions.

Legal Status: Furbearer.

Gray Wolf
Canis lupus

Field ID: This largest member of the dog family has a coarse coat of grizzled gray with silver, black, and reddish hairs intermixed. The body is rangy and muscular, the legs long, the face doglike with upright ears, and the tail long and bushy.

Size: Head/body length 43–48 inches (109–122 cm), tail 12–19 inches (30–48 cm), height at shoulder 26–28 inches (66–71 cm), weight 70–120 lbs (32–54 kg).

Habitat: Grasslands, open woodlands, mountain parks.

Distribution: Extirpated, but formerly statewide.

Field Notes: Wolves are highly social animals that hunt cooperatively, running down and killing prey, including deer, elk, bighorn sheep, and other large mammals. Packs range in size from two to eight, but have been known to number up to thirty-six animals. Unlike foxes and coyotes, wolves eat very little plant matter. Wolves maintain large home ranges and wander many miles in search of prey. The pack members communicate with facial expression, scent, barks, whines, howls, and tail and body posture. An alpha male and female are the dominant members of the pack and the only pair that breeds. A whelping den is excavated, or an existing burrow of another animal enlarged, and the female bears a litter averaging six puppies. All pack members aid in feeding the nursing female and rearing the pups. The pack may move the puppies one or more times. Pups don't mature until two years old and don't breed for another year. Eventually they're ejected by the adults, or leave to establish their own packs or join another. Gray wolves once roamed throughout Colorado but were extirpated by 1935, the subject of intense trapping, poisoning, and shooting efforts because of concerns about their predation on livestock, as well as long-held societal fears of wolves. Individual wolves have occasionally wandered into Colorado since restoration of the species to the Yellowstone region. The lack of this important predator impacts Colorado's natural ecosystem, but restoration, or even tolerance of natural expansion into Colorado, is controversial because of landowner concerns about livestock predation. In spite of societal fears and folklore, there are no confirmed records of attacks of healthy wolves on humans. The state wildlife commission is on record as treating wolves that arrive in Colorado on their own according to federal law, that is, as an endangered species. Colorado Parks and Wildlife provides updates on the status of wolves at http://wildlife.state.co.us.

Legal Status: State and federal endangered.

Kit Fox
Vulpes macrotis

Field ID: This small, delicate fox is pale, yellowish gray intermixed with black hairs, with yellow to white underparts. It has large ears, a small, pointed face with black markings, and a bushy tail tipped with black.

Size: Head/body length 15–20 inches (38–51 cm), tail 9–12 inches (23–30 cm), weight 3–6 lbs (1.4–2.7 kg).

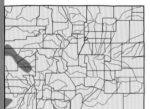

Habitat: Semidesert shrublands, piñon-juniper woodlands.

Distribution: Low elevations in southwestern Colorado and the lower Gunnison and Grand valleys along western parts of the Gunnison River.

Field Notes: The kit fox is the smallest fox in North America, a delicate-looking animal with very large ears that help it detect the movements of its prey. It is a nocturnal hunter, feeding largely on jackrabbits and kangaroo rats as well as birds, reptiles, and insects. Kit foxes inhabit arid, western landscapes. They form strong pair bonds and may mate for life. The foxes pair up in late fall and breed December through February. Kits are born February through March and average four to five in a litter. The adults dig multiple dens, moving their pups among the various sites. The male feeds the nursing female, then both adults hunt for the family, which disperses in late fall. Kit fox populations have declined drastically, primarily as nontarget victims of poisoning intended for coyotes, and the species is classified as endangered in Colorado. Some biologists have considered kit and swift foxes to be one species, but in Colorado, the two are distinctive in appearance and their ranges are separated by the Rocky Mountains.

Legal Status: State endangered.

What Color Is a Red Fox?

What color is a red fox? What about a black bear? Yes, these are trick questions. The common names of animals can be misleading. Most red foxes are red. A few are reddish with a dark cross covering the shoulders and the middle of the back. Some red foxes, though, are black with silver tips to their fur. They're called silver foxes.

That brings us to black bears. Many are not black. They may be dark brown to cinnamon to pale brown, or even blond. Some are sun-bleached blond on their backs with dark brown legs.

Swift Fox

Vulpes velox

Field ID: This small, slender fox is pale to yellowish gray intermixed with darker hairs, with paler underparts. It has black marks on the sides of the muzzle and a bushy tail tipped with black. The face is wide and triangular and the eyes narrow.

Size: Head/body length 15–20 inches (38–51 cm), tail 9–12 inches (23–30 cm), weight 4–6 lbs (2–3 kg).

Habitat: Grasslands.

Distribution: Eastern Colorado.

Field Notes: These Eastern Plains dwellers inhabit gently rolling grasslands, digging dens in the prairie soil, usually on a rise where they have a good outlook. While other foxes are fairly omnivorous, swift foxes eat mainly jackrabbits, prairie dogs, small rodents, birds, and other animal prey. They are nocturnal hunters. A mated pair forms a strong pair bond and excavates a den. The den may have up to six entrances, marked by bones, scraps of fur, and other food debris. Pups are born between March and May and average four to five in a litter. The male feeds the female with nursing young and helps feed the weaned pups. The family disperses in September and October. Swift fox populations declined drastically in the twentieth century and were extirpated over much of the Great Plains. They are common again in parts of eastern Colorado. Trapped for their furs, many are also killed as nontarget victims of poisoning and trapping aimed at coyotes. Swift foxes are less wary than other foxes and are easily trapped.

Legal Status: Species of special concern.

Red Fox
Vulpes vulpes

Field ID: These common foxes are most often orange-red with pale undersides, black legs, pointed ears, and a thick, bushy, white-tipped tail that is three-quarters the length of the body. Some have a dark cross over the shoulders (so-called cross foxes) or are black with silver tips on the hairs (silver foxes).

Size: Head/body length 22–25 inches (56–63 cm), tail 14–16 inches (36–41 cm), weight 10–15 lbs (5–7 kg).

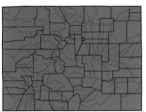

Habitat: Meadows; open woodlands; farm fields; pastures; riparian areas; urban, suburban, and rural parks; golf courses; greenbelts; and cemeteries.

Distribution: Statewide.

Field Notes: The adaptability and intelligence of red foxes is the stuff of folklore. They adapt well to life around people and also inhabit wilder habitats, usually near water. They are mainly nocturnal or crepuscular and need sufficient undergrowth to move around under cover. A mated pair moves into an existing burrow or digs a new one. The female bears four to five pups (also called cubs or kits) in early spring. Both male and female hunt for the family, with occasional help from immature females born the previous year. The young disperse in fall. The male and female usually stay together through winter and raise another family in spring. Foxes are omnivores, feeding on mice, birds and their eggs, berries, fruit, fish, carrion and insects. In spring, denning foxes produce a strong, musky (sometimes called foxy) odor. The adults will yip, bark, and yowl if intruders approach the den. An active den is conspicuous from the collection of bones, fish heads, fur scraps, and other debris around the entrance.

Legal Status: Furbearer.

Gray Fox
Urocyon cinereoargenteus

Field ID: This small, slender fox is salt-and-pepper gray with red ears, neck, and legs and white throat, chest, and belly. The very bushy, black-tipped tail is nearly as long as the body and has a black stripe of stiff hairs along the top.

Size: Head/body length 21–29 inches (53–74 cm), tail 11–16 inches (28–41 cm), weight 7–13 lbs (3–6 kg).

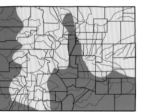

Habitat: Broken terrain with shrub-lands or piñon-juniper woodlands and riparian areas or edges of agricultural lands.

Distribution: Southeastern and north-ward along the foothills of the Front Range, far western Colorado.

Field Notes: The gray fox can be distinguished from the red fox by its color, shyer demeanor, and by the black, rather than white, tail tip. It does not adapt to life around people nearly as well as the red fox. Shy, secretive, and nocturnal, it emerges from its den in a hollow log, rock pile, or burrow to hunt birds and small mammals or forage for eggs, berries, fruit, carrion, and insects. Gray foxes are famous for their tree-climbing ability, grasping the trunk with the front legs and scrabbling upward with the hind feet. They are sometimes nicknamed "tree fox" and will climb after prey or to rest on a tree limb or crotch. A mated pair forms a pair bond in late winter to early spring. They den in burrows, rock piles, or even in hollow trees. A litter averaging four kits is born between March and May. The male feeds the female while she is nursing. The family hunts together through summer and disperses in the fall. The male and female often mate again the next year.

Legal Status: Furbearer.

Family Ursidae—Bears

Bears won't be mistaken for any other animal. They are very large, with a blocky shape and long legs. They have large heads; small eyes; small to medium, rounded ears; short tails; long muzzles; and a thick, coarse coat. They will rise on their hind legs but primarily move on four legs, walking on the soles of their feet (plantigrade). Though carnivores, with all the tools and adaptations of meat-eaters, bears eat a great deal of vegetation, up to 90 percent of their diet at certain times of year. Much of the flesh they eat is invertebrates and carrion.

Animal Sign: Large, five-toed, plantigrade tracks; ropelike or rounded piles of scat often containing berries and nuts; rubbed trees (in spring) with bark rubbed off on one side and dark, coarse hairs in the sap; day beds of leaves, needles, and plants scraped up into a circular mound; large boulders rolled over and rotting logs ripped open (seeking grubs); claw marks on trees showing five long, parallel scratches.

Black Bear
Ursus americanus

Family Ursidae—Bears

Field ID: This large, stocky animal ranges from pale blond to cinnamon to black. It has round ears, a boxy head with long muzzle, short tail, long legs, and five-toed paws.

Size: Head/body length 5–6 feet (1.5–1.8 m), tail 3–5 inches (76–127 mm), height at shoulder 2–3 feet (61–91 cm); weight, male up to 300–500 lbs (136–227 kg), female 88–155 lbs (40–70 kg).

Habitat: Piñon-juniper woodlands, mountain meadows, forests, riparian areas, and shrublands.

Distribution: From eastern foothills west throughout central and western Colorado.

Field Notes: Black bears live in nearly any habitat with mature stands of oakbrush, chokecherry, and other fruit and nut-producing shrubs. Despite their fearsome reputation as predators, up to 90 percent of their diet is plant material. They eat acorns, pine nuts, berries, roots, flowers, carrion, small mammals, fawns, birds and their eggs, insects, grubs, and garbage (when careless humans invite them to it). Bears are diurnal or crepuscular except in late summer and fall, when they spend up to 23 hours a day feeding. They must store enough energy to last through winter hibernation, when they may go for up to six months without eating or drinking. Bears begin entering winter dens in early October but may den as late as early December. Their dens are within rock piles or excavated under shrubs or trees. Some bears

188

just crawl under some cover and hibernate on the ground. Many dens are used in successive years. Bears emerge in spring in March or April, having lost up to 25 percent of their autumn body weight. For a few weeks, as their bodies rev back up, they move around in a sort of walking hibernation, eating and drinking very little. Black bears breed in early June through mid-August but don't form a pair bond. Females give birth during hibernation in late January and February to usually two—sometimes three—tiny, very immature babies that each weigh only half a pound. The cubs stay with the mother through the first summer and then hibernate with her. When they emerge in spring, she drives the yearlings away and mates again. Such large animals are slow to mature, and females don't generally bear young until they are five years old. They have cubs only every other year. Black bears can be distinguished from grizzlies by their straight, rather than dished-in, snout, larger ears, hind legs higher than shoulders, lack of a shoulder hump, lack of grizzled hairs or a mane of hair over the shoulders, and no claws showing in their track. Bears walk on the soles of their feet, so their track shows a large main pad (sole of the foot) and five toes.

Legal Status: Big game.

Grizzly Bear
Ursus arctos

Field ID: This very large bear is pale to dark brown with silvery guard hairs, a mane of long hair over the shoulders, and a prominent shoulder hump. The ears are small and rounded, the face dish-shaped, the eyes small. The grizzly has long legs and large, five-toed paws.

Size: Head/body length 6–7 feet (1.8–2.1 m), tail 3 inches (76 mm), height at shoulder 3–3½ feet (91–107 cm), weight 300–600 lbs (136–272 kg).

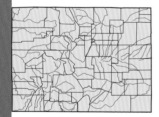

Habitat: Prairies, mountain meadows, forests, shrublands, alpine tundra.

Distribution: Extirpated, but formerly statewide. No grizzly bears are known to currently inhabit Colorado.

Field Notes: People are surprised to know that grizzly bears, now limited to remote mountains in the northern Rockies, Canada, and Alaska, once lived on the prairie. Grizzly bears are significantly larger than black bears. They have a pigeon-toed, five-toed track that shows their long claws. Both bear species have a similar life history, emerging from hibernation in spring (the adult females with one to three young), feeding through summer and reentering hibernation in the fall. Grizzlies have a voracious appetite and will eat anything from live prey to berries to carrion. Preying on livestock created considerable conflicts with stockmen, and the bears were heavily trapped, poisoned, and shot through the mid-twentieth century. Grizzlies were thought to be extirpated from Colorado until one was killed in the San Juan Mountains in 1979. There continue to be occasional unconfirmed sightings. The absence of the grizzly bear leaves a significant void in Colorado's natural ecosystem, but it is unlikely the great bear will ever be tolerated again in the state, especially as the human population continues to move into once-wild lands. Any sightings or likely sign should be reported to Colorado Parks and Wildlife.

Legal Status: State endangered, federal threatened.

Two Strikes and They're Out

Bears have an extremely acute sense of smell and are always in search of food. Garbage, bird feeders, or barbecue grills left around a house or campsite in bear country will likely attract them. Once a bear has found food at a location, it will return to that spot seeking a meal. That's when bears get into trouble. Colorado Parks and Wildlife will trap and move a problem bear after the first complaint. But under a rule nicknamed "Two Strikes, You're Out," wildlife officers must kill a bear that bothers people more than once. Anyone living or recreating in bear country can help save the life of a bear by never leaving out anything—from dog food to greasy grills—that might attract a bear. Trash should be stored indoors or in a bear-proof can, pets should be fed indoors, and bird feeders either eliminated or brought in at night.

The Colorado Parks and Wildlife website, http//wildlife.state.co.us, has a great deal of information on reducing conflicts with bears.

Family Procyonidae—Raccoons and Relatives

Of the three species of this family found in North America, two inhabit Colorado—the raccoon and ringtail. (The third, the coati, is found in parts of southern Texas, New Mexico, and Arizona.)

Procyonids have pointed muzzles; faces marked in black or white; small to medium rounded ears; and long, very bushy tails marked with dark rings from base to tip. The raccoon is stocky while the ringtail is slender. Males are larger than females. They have five toes on each foot and walk flat-footed (plantigrade).

Though classified as carnivores, in their food habits they are omnivorous, with the plant matter they feed on being mainly berries, nuts, fruits, and seeds. Both raccoons and ringtails are nocturnal or crepuscular.

Animal Sign: Plantigrade tracks in mud like a tiny bear's, showing five toes, the sole of a foot, and the palm of a "hand."

Ringtail
Bassariscus astutus

Field ID: This cat-sized carnivore has a long, slender body and extremely bushy tail that is as long as the head and body. It is yellowish gray with a pointed muzzle, large eyes ringed with white, pointed, upright ears, and black and white rings down the tail.

Size: Head/body length 14–16 inches (36–41 cm), tail 15 inches (38 cm), weight 2–2½ lbs (907–1134g).

Habitat: Canyons, cliffs, rocky hillsides of piñon-juniper woodlands, foothills shrublands, and pine-oak forests.

Distribution: Across the western two-thirds of the state and southeastern Colorado.

Field Notes: Ringtail cat, miner's cat, civet cat—the nicknames for the ringtail are many. Ringtails do look like a mix of animals with their catlike body, foxy face and raccoon tail. Shy, nocturnal hunters, ringtails became familiar to miners during Colorado's hard-rock mining heyday. Miners welcomed them around their cabins once they realized how many mice and woodrats one ringtail could kill. They learned to recognize ringtails in the dark from the red to yellow-green shine of their eyes. Ringtails are agile, nimble climbers, able to rotate their hind feet 180 degrees, leap, ricochet, and chimney-climb as they move around rugged canyons and cliffs. They have semiretractile claws. Ringtails eat almost any available food, hunting mice, woodrats, lizards, bats, and insects and feeding on fruits, berries, juniper berries, and carrion. When threatened, ringtails give a fox-like bark and also whimper and growl. They den in rock crevices, caves, hollow trees, mines, and old buildings. The female bears one litter a year of three to four young, which go out on their own within five months.

Legal Status: Furbearer.

Raccoon
Procyon lotor

Field ID: This medium-sized animal is stocky, with short legs, rounded ears, and a long, bushy tail marked with black rings. The gray fur is tipped with black, and there is a black mask across the eyes.

Size: Head/body length 18–28 inches (46–71 cm), tail 8–12 inches (20–30 cm), weight 12–35 lbs (5–16 kg).

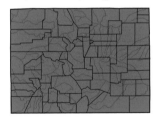

Habitat: Streams, canals, ponds, lakes, riparian areas, farmland, suburban parks, and gardens.

Distribution: Statewide up to about 10,000 feet.

Field Notes: Raccoons are among the most familiar animals in North America. Not common in Colorado prior to settlement, raccoons have expanded into suitable habitat statewide. They are most common along lower elevation streams and ponds with riparian cover of shrubs and trees. These highly adaptable, nocturnal foragers feed on crayfish, fish, insects, larvae, nuts, seeds, berries, crops, garbage, eggs, small mammals, and carrion. The characteristic face mask and cryptic patterns of the fur help camouflage the raccoon at night. Though humans may not see their raccoon neighbors, they are likely to find their planti-grade tracks in the mud along a stream, the front tracks shaped like tiny hands and the back like tiny bear or human feet. Raccoons are active year-round and are usually solitary, except for a mother with her young, though in winter they may gather in groups of up to two dozen animals. Young are born in the spring, with litters usually of two to five. They are weaned within about three months.

Legal Status: Furbearer.

195

Family Mustelidae—Weasels

Members of the weasel family vary greatly in color and size, from the 1-ounce female short-tailed weasel to the 60-pound male wolverine. But they are fairly consistent in basic body shape. They mostly have long, "weasely" bodies; short legs; long, fairly bushy tails; long necks; small heads; pointed muzzles; large eyes; and small, rounded ears. Their front and hind feet have five toes. Males are larger than females. Mustelids are adapted for various lifestyles. Martens climb trees after prey, ferrets and badgers hunt underground, mink and otters hunt fish and aquatic prey in the water, and weasels hunt on the ground. They are inquisitive and known for strength and ferocity, often killing prey larger than themselves. After they seize a prey animal, weasels may wrap their long, muscular bodies around it, sinking their sharp teeth into the neck or skull. They squeeze and strangle large prey as well as biting it.

Anal musk glands are a defining trait of this group. They are crepuscular or nocturnal, though sometimes active during the day. In many mustelids, the implantation of embryos is delayed for several months after mating.

Animal Sign: Five-toed track showing toes spread in a 1-3-1 pattern; bounding track measuring 12 to 48 inches between leaps; tracks in snow may disappear—as the animal dives beneath the snow—and reappear further on.

Pine (American) Marten
Martes americana

Field ID: This large weasel is golden-brown to dark brown with paler underparts; a buff throat; upright, rounded ears and a pointed muzzle. The tail is long and bushy and half the length of the head and body.

Size: Head/body length, male 16–17 inches (41–43 cm), female 14–15 inches (36–38 cm); tail, male 8–9 inches (20–23 cm), female 7–8 inches (18–20 cm); weight, male 1²⁄₃–2¾ lbs (754–1248 g), female 1½–1⅞ lbs (681–851 g).

Habitat: Lodgepole pine forests, spruce-fir forests, alpine tundra.

Distribution: Through higher elevations of the central mountains.

Field Notes: The pine marten is an animal of mountain coniferous forest. It prefers forests with a tree cover of about 50 percent to provide sufficient cover, rest sites, and hunting and foraging grounds. Old growth forests offer ideal habitat because of dense undergrowth, deadfall, and standing dead trees. Martens hunt voles, mice, pine squirrels and ground squirrels, snowshoe hares, chipmunks, shrews, and other prey and feed on the berries of mountain shrubs, carrion, and other available food. Their lithe bodies are adapted to pursue prey along the forest floor and up trees. They are agile climbers, with semiretractile claws. Martens are active year-round and, though mainly crepuscular and nocturnal, are sometimes spotted by hikers and skiers during the day, resting on tree limbs. Their size, alert posture, and pale throats are helpful in identification. Martens den in hollow trees, rock piles, and burrows. They breed from late July through September, producing a single litter a year averaging three young.

Legal Status: Furbearer.

Short-tailed Weasel (Ermine)
Mustela erminea

Field ID: This small weasel has a long, slender body, short legs, and a short, black-tipped tail. In summer, it is pale brown to chocolate brown, with buff to white underparts, turning pure white in winter except for the tip of the tail.

Size: Head/body length, male 6–9 inches (15–23 cm), female 5–7½ inches (13–19 cm); tail, male 2¼–4 inches (57–102 mm), female 2–3 inches (51–76 mm); weight, male 2½–6 oz (71–170 g), female 1–3 oz (28–85 g).

Habitat: Aspen forests, subalpine forests, alpine tundra, and talus slopes.

Distribution: Throughout the central mountains of the state, from 6,000 to 14,000 feet

Field Notes: This weasel, the smallest carnivore in Colorado, is mainly nocturnal. It alternates between busy foraging bouts of about an hour and rest periods. It is active year-round, preying on voles, mice, shrews, and small mammals. The paws are thickly furred between the pads. Short-tailed weasels are busy, active hunters, scurrying along the forest floor and easily climbing trees. They can kill prey much larger than they are, such as young rabbits. Excess food is cached. Like snowshoe hares and ptarmigan, they undergo a dramatic molt from brown in summer to snowy white in winter. During the fall and spring transitions, they are a mottled brown and white. Males are significantly larger than females. A single litter of six to nine young is born between May and July. The female cares for the kits. Also known as the ermine, the short-tailed weasel ranges from the northern United States through Canada into northern Eurasia, where its black-tipped, white tail was traditionally used to decorate the cloaks of royalty.

Legal Status: Furbearer.

Long-tailed Weasel
Mustela frenata

Field ID: This large weasel has a long, slender body, short legs, long tail, short muzzle, and upright, rounded ears. It ranges from light to cinnamon brown, with a white or yellowish throat, chest, and undersides. In winter, it molts to pure white, with a black-tipped tail. In southeastern Colorado, it may have a white mask or bridle pattern on the face.

Size: Head/body length, male 9–10½ inches (23–27 cm), female 8–9 inches (20–23 cm); tail, male 4–6 inches (102–152 mm), female 3–5 inches (76–127 mm); weight, male 7–12 oz (198–340 g), female 3–7 oz (85–198 g).

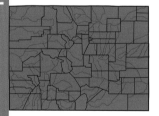

Habitat: Grasslands, shrublands, riparian areas, forests, agricultural lands, suburban areas.

Distribution: Statewide.

Field Notes: Through most of Colorado, if you see a weasel, it is likely to be this one. It is found in nearly every habitat in the state, usually near water. These busy carnivores are active day and night, year-round. When hunting or fleeing danger, they run with their long tails held straight up. They dart down a hole or into cover, then pop their heads up curiously to look around. Skiers may see them dive into the snow and pop up again further along, discernible against the snow by their black eyes and tail tip. When running, they typically bring the hind legs forward before stretching out with the forelegs, resulting in a hump-backed, inch-worm-style run. They mainly hunt small mammals. The smaller females, able to dart down burrows, probably feed more on mice and voles while the larger-bodied males hunt larger prey such as cottontails. They typically kill their prey with a bite to the back of the neck at the base of the skull. These weasels also feed on birds, insects, fruits, and berries. They are known for being noisy and quarrelsome, challenging other weasels that enter their territory with high-pitched squeaks and growling. Long-tailed weasels build nests of grass and the fur from their prey, making use of the burrows of other animals. They breed in July and August and bear one litter in April or May averaging seven young.

Legal Status: Furbearer.

Black-footed Ferret
Mustela nigripes

Field ID: This large weasel has a long, slender body; short legs; short, rounded ears; and a long, black-tipped tail. It is buffy to tan, with darker brown down the head and back, a white throat, dark brown legs, and a black mask across the eyes. It does not change color in winter.

Size: Head/body length 15–18 inches (38–46 cm), tail 5–6 inches (13–15 cm), weight 1½–2 lbs (680–907 g).

Habitat: Prairie dog towns in grasslands and shrublands.

Distribution: Formerly nearly statewide; being restored in northwestern Colorado.

Field Notes: Black-footed ferrets live in prairie dog towns and are dependent on prairie dogs as their main food. They are nocturnal, resting in abandoned prairie dog burrows during the day and emerging after dark to prey on their hosts. Their long, slender bodies and short legs are ideally suited for moving down burrows. Because of their scarcity and shy nature, not a lot is known about the species' life history in the wild. Captive animals breed in March and April. A single litter is born in May, averaging three to four young. Black-footed ferrets are a different species than the European ferret that is kept as a pet. Though never numerous, black-footed ferrets were probably once widespread, their range mirroring that of prairie dogs, but they were extirpated from Colorado by the mid-twentieth century with the eradication and reduction of prairie dog populations. Beginning in 2001, biologists with the Colorado Division of Wildlife (now Colorado Parks and Wildlife) and the US Fish and Wildlife Service released captive-bred (and in 2004 wild-born) ferrets into white-tailed prairie dog colonies in northwestern Colorado. Colonies are now established at Coyote Basin west of Rangely and at the Bureau of Land Management's Wolf Creek Management Area southeast of Dinosaur National Monument. Any sightings of black-footed ferrets should be reported to Colorado Parks and Wildlife.

Legal Status: State and federal endangered.

The Rarest of Mammals

The black-footed ferret is one of the rarest mammals in North America. Shy of habit and nocturnal—a Pawnee legend describes them as "staying hid all the time"—ferrets went quietly about their lives little noticed. As prairie dog populations declined with the spread of agriculture, black-footed ferrets, which depend on prairie dogs for food and cover, all but disappeared.

By 1979 the black-footed ferret was thought to be extinct. Then in 1981, a Wyoming rancher's dog brought home the fresh-killed carcass of a ferret. A surviving population of 129 ferrets was discovered. Canine distemper reduced that group to fewer than 20 animals, which were brought into a captive breeding program.

Decades of careful breeding and planned releases into the wild have re-established closely monitored populations in Colorado, Arizona, Montana, New Mexico, South Dakota, Utah, and Mexico. Black-footed ferrets are still one of the rarest mammal species in North America, and it's too soon to know if they can eventually survive, unmanaged, in the wild. The US Fish and Wildlife Service Black-footed Ferret Conservation Center is located in Fort Collins. For more information, visit www.blackfootedferret. org. You can see captive black-footed ferrets at the Cheyenne Mountain Zoo in Colorado Springs.

Family Mustelidae—Weasels

Mink
Neovison vison

Field ID: This medium-sized weasel has a thick coat of soft, glossy fur with long guard hairs. The fur is chestnut to very dark brown. It has a long, slender body; short legs; small, round ears; short muzzle; partially webbed paws; and a long, furry tail.

Size: Head/body length, male 13–17 inches (33–43 cm), female 12–14 inches (30–36 cm); tail, male 7–9 inches (18–23 cm), female 5–8 inches (13–20 cm); weight, male 1½–3 lbs (680–1361 g), female 1¼–2⅖ lbs (567–1089 g).

Habitat: Wetlands, marshes, ponds and streams, riparian areas.

Distribution: Statewide.

Field Notes: Mink stink. They are the smelliest of the generally smelly weasels, though not as aromatic as skunks. Mink release a strong-smelling liquid from their anal glands when threatened, though they can't aim and fire like a skunk can. Mink are animals of wetlands, where they prey on crayfish, muskrats, fish, mice, snakes, frogs, ground-nesting birds, and eggs. They build dens alongside or in the banks of streams and ponds, or within old beaver lodges or muskrat dens. They are very good swimmers, favoring ponds and waterways with a good cover of willows and other vegetation. Mink are nocturnal in the warm months, becoming diurnal during winter. They don't hibernate but may rest in their dens for days at a time during bad winter weather. Mink are solitary animals except during the breeding season in late winter. Females have one litter of about five young, who stay with her until fall. Littermates may stay together when they disperse until they find their own wintering sites. Mink are valuable furbearers, and humans are their primary predator.

Legal Status: Furbearer.

Wolverine
Gulo gulo

Field ID: These very large, blocky-bodied weasels have longish legs, a bushy tail about one-fourth the length of the body, a broad head, a short, thick neck, and short, round ears. Their heavy coat is medium to dark brown, with broad, yellow stripes that run from the shoulders to the rump.

Size: Head/body length 29–32 inches (74–81 cm), tail 7–9 inches (18–23 cm), weight 35–60 lbs (16–27 kg).

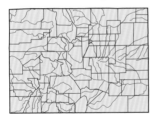

Habitat: Upper montane and subalpine forest, alpine tundra.

Distribution: Potentially at higher elevations through the central mountains, especially where snow lies deep well into spring. No wolverines are known to currently inhabit Colorado.

Field Notes: The wolverine looks more like a small bear than a weasel. Its body is stocky and muscular rather than long and slender, and it is significantly larger and heavier than its weasel cousins. Its Latin name translates to "gluttonous glutton," an apt description of its aggressive hunting style and voracious appetite. The wolverine has a reputation for ferocity. It's very strong for its size and will kill prey much larger than itself. Most of its prey are small mammals, birds, eggs, and fish. It also eats carrion, roots, and berries. The wolverine was thought to be extirpated from Colorado, the last confirmed sighting being in 1919. But in 2009 a radio-collared wolverine was tracked into north central Colorado from the Yellowstone area, the first confirmation of the species in the state in 90 years. Wolverines were probably never numerous in Colorado.

They have declined throughout their range from trapping and habitat fragmentation. They are active year-round, mostly at night. Wolverines breed in summer but delay implantation, and a single litter of one to four young is born in spring.

Legal Status: State endangered.

American Badger
Taxidea taxus

Field ID: These large members of the weasel family are a grizzled, silvery-gray with long, white-tipped guard hairs. They have short legs, long front claws, a short, bushy tail, rounded ears, and a flat head. The feet and lower legs are black, and the black face is marked with white, including a distinctive white stripe from the nose across the top of the head.

Size: Head/body length 18–22 inches (46–56 cm), tail 4–6 inches (102–152 mm), weight 13–25 lbs (6–11 kg).

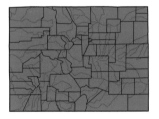

Habitat: Grasslands, forest edges, shrublands, mountain meadows, alpine tundra, open mountain forests.

Distribution: Statewide.

Field Notes: The badger is a digging machine, with powerful legs and long, strong front claws (up to 2 inches long) that can rip into hard-baked prairie soil and tunnel quickly after burrowing prey. Badgers are found in open habitats statewide, especially where there are good populations of prairie dogs and ground squirrels. They are opportunistic hunters and will eat any small mammals, lizards, snakes, and ground-nesting birds they can catch. Aggressive and well equipped to defend themselves, badgers will face off against an attacker by backing down a burrow entrance and presenting snarling teeth and ferocious claws to the enemy. Collisions with vehicles and intentional killing by humans are the main causes of death. An Idaho study found that 60 percent of known badger mortalities were human-caused. Badgers are solitary except during the breeding season. They mate in late summer but don't have a high reproductive rate. Implantation of embryos is delayed and females produce one litter, averaging two young, in March or April. By fall, the young disperse to live on their own.

Legal Status: Furbearer.

Northern River Otter
Lontra canadensis

Field ID: This weasel relative has a long, muscular body with a long, thick tail and smooth, sleek coat. The small, flat head has long whiskers and small round ears. The legs are short, the paws large and webbed. The hind feet have rough pads for gripping slippery streambeds and banks.

Size: Head/body length 26–30 inches (66–76 cm), tail 12–17 inches (30–43 cm), weight 10–25 lbs (5–11 kg).

Habitat: Riparian areas along streams and rivers through shrublands and forests from foothills to subalpine elevations.

Distribution: Along the upper Colorado, Gunnison, Dolores, Piedra, Laramie, and Cache la Poudre rivers.

Field Notes: River otters are highly mobile animals, traveling stream and river systems through a variety of habitats and elevations. Fish are their principal food; they also eat crayfish, frogs, small mammals, and birds. Otters are active year-round, being mainly nocturnal in summer and becoming more diurnal in winter. They use beaver lodges, logjams, riparian thickets, and snow and ice caves as den sites. Otters and mink are similar in appearance and often mistaken for each other, but otters are much bigger. They are social animals living in family groups. The group, led by an adult female, is comprised of yearlings, juveniles, and other adults, but males don't regularly stay with the family. Otters breed in spring, but implantation is delayed for months, until mid-winter. A single litter, usually twins, is born in March or April and females mate again while still nursing their kits. Once found along rivers through much of the state, otters were heavily trapped for their rich fur and were extirpated from Colorado. Restoration efforts, begun in 1976, have returned otters to some mountain river systems but otters are still neither numerous nor widespread. Report any sightings to Colorado Parks and Wildlife.

Legal Status: State threatened.

Family Mephitidae—Skunks

Long classified as mustelids, new genetic information shows that skunks are different enough from weasels to be classified in their own family. The family name Mephitidae comes from a Latin term meaning (surprise, surprise) "bad odor." The modified anal glands that produce the noxious defensive spray are a defining characteristic of this family. Skunks are also dramatically patterned and colored in black and white as a visual warning (and reminder for the previously skunked!) to animals that might threaten them.

Skunks have blocky bodies; long, very bushy tails; small eyes; and five toes on each foot. They are nocturnal and omnivorous, trundling along and foraging for anything edible, turning over logs, digging into the ground or into burrows after prey, eating whatever berries, nuts, roots, grubs, carrion, or prey they can find.

Animal Sign: A foul-smelling, skunky odor; small depressions dug in the ground; overturned logs and rocks; five-toed, plantigrade tracks with a double foot pad.

Western Spotted Skunk
Spilogale gracilis

Field ID: This small skunk is dark brown to black, with rather narrow, somewhat irregular white stripes along the head, back, and sides that are broken into spots. There is a white spot between the eyes and under each ear. The plumed tail is white on the tip and along half of the underside.

Size: Head/body length 9–12½ inches (23–32 cm), tail 4⅓–6¼ inches (110–159 mm), weight ⅞–1½ lbs (397–680 g).

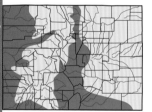

Habitat: Rocky areas of shrublands, montane forests and piñon-juniper woodlands.

Distribution: Lower elevations in the central mountains and canyon country of western Colorado.

Field Notes: Spotted skunks are significantly smaller than the more-familiar striped skunks. The western spotted skunk is an animal of dry, rocky landscapes and climbs well through rugged terrain. Spotted skunks feed on insects, spiders, small mammals, birds, bird eggs, fruits, berries, and carrion. Skunks are nocturnal, foraging from sunset to sunrise. They are solitary animals, though females may den together in winter. Mating is in fall, but implantation is delayed, and the female bears about four young in late spring. Spotted skunks are active year-round, building dens among the rocks, in crevices and under sheds and abandoned buildings. When threatened, they raise their hindquarters into the air in a "handstand" and point the anal glands at the threat. The pattern of white on black varies on individual animals, but the characteristic broken-stripe spotting on black fur is particular to this species.

212   **Legal Status:** Furbearer.

Eastern Spotted Skunk
Spilogale putorius

Field ID: This small skunk has black fur with four to six white stripes broken into spots down its back and sides. There is a white spot between the eyes and the tail tip is black.

Size: Head/body length 10½–12 inches (27–30 cm); tail 5–11 inches (13–28 cm); weight, male 1–2 lbs (454–885 g), female ½–1 lb (227–454 g).

Habitat: Farm fields and agricultural areas, riparian areas, shrublands, and grasslands near water.

Distribution: Eastern Colorado, along the South Platte and Arkansas rivers.

Field Notes: Eastern and western spotted skunks were once considered one species. The eastern form is larger than the western, and their ranges in the state do not overlap. The eastern spotted skunk is rare in Colorado, with only a few records from the state. It is an animal of open woodlands and agricultural land. It dens in the burrows of ground squirrels, gophers, and weasels, in caves, under sheds and woodpiles, even in abandoned vehicles. Eastern spotted skunks breed in spring and bear a single litter of four or five young in June. Like other skunks, they are omnivorous, with insects particularly important in their diet. They also eat small mammals, fruit, and the field gleanings of crops such as corn and wheat. Spotted skunks don't hibernate, though they gather in dens in the winter, when their activity is reduced. The tails of spotted skunks are shorter than those of other skunks. Their defensive spray is more acrid than the striped skunk's.

Legal Status: Nongame.

Striped Skunk
Mephitis mephitis

Field ID: This large skunk has glossy black fur with a large white stripe beginning at the base of the skull, splitting into two stripes that travel along the back and converge again at the tail. There is a strip of white on the nose. The long, plumy tail is black interspersed with white.

Size: Head/body length 13–18 inches (33–46 cm), tail 7–10 inches (18–25 cm), weight 6–14 lbs (3–6 kg).

Habitat: Agricultural land, towns, cities, suburbs, riparian areas, meadows, shrublands, open woodlands and forest edges, all generally near water.

Distribution: Statewide up to 10,000 feet.

Field Notes: Leaving camouflage to other animals, the skunk's black-and-white coat shouts a warning. Making creative use of its musk glands, the skunk comes armed with an oily, foul-smelling defensive spray that deters most predators, including humans. When threatened, the skunk will stamp its front feet, arch its back, raise its tail, shuffle backward, and finally let loose with a blast of noxious spray. Great horned owls, however, have a poor sense of smell and are known to prey on skunks. Vehicles are also undeterred by the skunk's warning colors, and being run over (along with other intentional human actions) is a major cause of skunk mortality. Skunks are omnivores, trundling around from dusk until dawn, pawing and snuffling in, under, and around burrows, logs, and rocks for grubs, insects, birds' eggs, small mammals, berries, nuts, or anything edible. They don't hibernate but become much less active in winter. They dig burrows or move into existing shelters, mating in February or March, producing a single litter of five to eight young in early summer. Males sometimes gather harems of females. The young stay with the mother until they are about three months old.

Legal Status: Furbearer.

Pee-yew!

Anyone who has tangled with a skunk does not quickly forget the experience. Once sprayed, few nonhuman predators will bother a skunk again, either.

The skunk's defensive spray is produced by musk glands that are typical of weasels and other members of the family Mustelidae, and, until recently, skunks were classified as mustelids. Skunks can control the release of this spray. Like cannoneers, they can fire their guns at will, sending as many as eight bursts as far as 12 feet before running out of ammunition. When a skunk feels threatened, it turns its hind end, the "business end," toward its enemy. The openings of the musk glands pop out, muscles contract, and the oily, yellowish musk fires out, in either a thick stream or a fine spray.

The skunk's "perfume" is the chemical compound butyl mercaptan. It is in the same family of sulfurous chemicals as the compound added to odorless natural gas to make it detectable. Skunk smell has been described as "a mixture of strong ammonia, essence of garlic, burning sulfur, a volume of sewer gas, a sulfuric acid spray, and a dash of perfume." This oily, overpowering musk can cause sneezing, coughing, choking, gagging, and vomiting. The caustic vapor burns eyes and nasal membranes. A victim who receives a dose full in the face can be temporarily blinded. Considering the sensitive noses of many predators, it's a pretty effective deterrent.

Family Felidae—Cats

Members of the cat family are highly specialized and wonderfully adapted for their role as predators. While others of the order Carnivora eat some to a lot of plant matter, cats eat almost none. They depend on live prey and rarely eat carrion. As consummate hunters, they have lithe, muscular bodies, long legs, upright ears, roundish heads, daggerlike canine teeth, and large, soft paws with very sharp, retractile claws. Their large eyes are set at the front of the skull, giving them binocular vision to aid in judging distance as they leap onto prey.

While pack hunters like wolves run down their prey, cats are solitary hunters that stalk and ambush, leaping down onto prey or rushing for a short distance and pouncing. Cats have a very large "gape"—their jaws can open almost 90 degrees. Their very sharp claws, usually retractable into sheaths, curve out and around to grasp and pierce their quarry. Because an injury can mean death for these solitary hunters, cats wait and watch for an optimal opportunity before attacking.

Cats have four toes on the hind foot and five toes on the front, but the fifth (the dewclaw) is up on the leg. They walk on the toes (digitigrade).

Animal Sign: Four-toed paw track with heel pad; claws rarely show in track; track generally wider than long; front paws larger than hind; prey cached beneath tree litter; scrapes of soil and litter with feces and urine; droppings similar to domestic cat (though larger) and often covered with soil.

Mountain Lion
Puma concolor

Field ID: This large cat is sleek and muscular, varying in color from grayish tan to reddish brown, with small ears, rounded head, short, boxy muzzle, and long, heavy tail that makes up one-third or more of the total length.

Size: Head/body length, male 3½–5 feet (107–152 cm), female 3–4 feet (91–122 cm); tail, male 30–36 inches (76–91 cm), female 10½–14½ inches (27–36 cm); height at shoulder 26–31 inches (66–79 cm); weight, male 80–264 lbs (36–120 kg), female 64–140 lbs (29–64 kg).

Habitat: Rough, broken terrain, especially canyons and foothills, with open woodlands.

Distribution: Nearly statewide but especially from Front Range foothills west across the state.

Field Notes: Deer are the mountain lion's main food, so where there are mule deer (in suitable lion habitat), there are likely to be lions. They also eat smaller live prey. Consummate predators, they are equipped with sharp, retractile claws that curve out to grab and hold prey; long, sharp teeth; swift, athletic bodies; and large, forward-facing eyes. They generally wait in ambush, preferring to leap down onto prey and kill with a bite to the back of the neck, or to rush and drag down prey. They will cache their kill under leaf litter or snow. Lions are solitary animals, except when the female has kittens (also called cubs). There is no specific breeding season, though mating is probably most common in spring, with kittens born in midsummer. A female bears one litter about every one-and-a-half to two years, usually of two to three kittens, which she rears by herself. Lions are active year-round, both day and night. They are highly mobile and can range widely. Though attacks on humans are rare, conflicts are increasing as people move into mountain lion habitat. People living in lion country can reduce the risk of conflicts by cutting back vegetation around their home to remove hiding places, feeding pets indoors, and never leaving any trash, food, or pet food outdoors that will attract small mammals, which might, in turn, attract lions.

Legal Status: Big game.

Canada Lynx
Lynx canadensis

Field ID: This medium-sized wildcat is pale, grizzled gray with flecks of black and sometimes a wash of red. It has long legs, a short tail with a complete black tip, triangular ears with pronounced furry tufts that curve conspicuously downward, and a ruff of fur around the face. The paws are very large.

Size: Head/body length 32–36 inches (81–91 cm), tail 4 inches (102 mm), weight 15–30 lbs (7–14 kg).

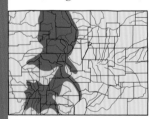

Habitat: High mountain coniferous forests, forest edges.

Distribution: The central mountains above about 9,000 feet.

Field Notes: Lynx are mainly animals of the northern boreal forests of Canada and Alaska. A slender finger of their range extends down the Rockies into Colorado. Lynx have probably never been common in Colorado. A restoration program begun in 1999 has returned lynx to many areas of the central mountains. The released animals have moved into a variety of habitats, and they feed on more diverse prey than predicted by biologists. Lynx look much like bobcats but are larger, have shorter tails, larger ear tufts, and less-patterned coats. They have very large paws, the better to disperse their weight in deep snow as they pursue their favorite food, the snowshoe hare. Hares comprise 80 percent of the lynx diet, and lynx distribution shadows that of snowshoe hares. Lynx populations are also tied to fluctuations of hare populations. An average of four kittens are born in April or May, but when hare numbers are low, the kittens don't survive. Lynx have bred successfully in the state since restoration and are thought to be expanding their range throughout the Colorado high country. Any sightings should be reported to Colorado Parks and Wildlife.

Legal Status: State endangered, federal threatened.

Bobcat
Lynx rufus

Field ID: This medium-sized cat has a grayish buff to reddish buff coat marked with darker spots and streaks, paler undersides and dark bands around the legs. It has a shortish tail with an incomplete black tip, long legs, a ruff of fur around the face and triangular ears with tufts of fur.

Size: Head/body length 25–30 inches (64–76 cm), tail 5 inches (13 cm), weight 15–35 lbs (7–16 kg).

Habitat: Piñon-juniper woodlands, low-elevation forests, rough foothills and canyons, shrublands.

Distribution: Statewide.

Field Notes: The bobcat is about twice the size of a domestic cat, with a somewhat-short, "bobbed" tail, long legs, a ruff of fur around the face and tufts of fur bristling from the tips of the triangular ears. Bobcats are the most abundant and widely distributed of Colorado's three cat species. They prefer broken country with good cover but are found in all habitats except open prairie, farmland, and urban areas. During the day, they rest in rocky or wooded areas with dense vertical cover but little ground vegetation so they can watch for prey without being seen. Bobcats are active year-round. They are mainly crepuscular but also nocturnal and diurnal, hunting rabbits, squirrels, mice, small birds, and the occasional deer. In summer the bobcat's coat is reddish, turning grayer in winter. Its Latin name, *Lynx rufus*, means "red wildcat." Bobcats are solitary animals, except when the female has a litter of spotted kittens. She bears three or four kittens in May or June and hunts for and raises them by herself.

Legal Status: Furbearer.

ORDER
PERISSODACTYLA

Odd-toed Hoofed Animals

The order Perissodactyla includes horses, zebras, asses, rhinos, and tapirs and is defined as hoofed animals with an odd number of toes. This toe count differs from the even number of toes found in the "cloven hoof" of deer, sheep, goats, cattle, and pronghorn.

Most perissodactyls have one to three functional toes on each foot, with the axis, and thus the weight balance-point, running through the middle digit. Horses, zebras, and asses, which all belong to the genus *Equus*, have a single toe—the familiar horse hoof. Rhinos have three toes. But tapirs have four toes on the front foot (with a vestigial fifth toe) and three toes on the hind foot.

All in this group have elongated skulls and graze or browse on coarse vegetation such as grass or leaves. Rather than the ruminating stomach of even-toed hoofed mammals (artiodactyls), perissodactyls have a simple but elongated digestive system.

Perissodactyls differ greatly in size, from the 600-pound tapir to the 5,000-pound white rhino. This once-large order now has only about 17 surviving species, compared to about 170 species of artiodactyls.

Family Equidae—Horses

This family includes only a single, surviving genus, *Equus*, which includes zebras, wild and domestic asses and donkeys, and wild and domestic horses. The Mongolian wild horse, or Przewalski's horse (*Equus ferus*), is the only surviving wild horse species. The wild horses of the American West are feral domestic horses. However, some taxonomists do not consider them as separate species. The fossil record shows that horses evolved on the open grasslands of North America, spread to Eurasia via the Bering Strait land bridge but subsequently died out in North America in the Pleistocene. Beginning in the sixteenth century, horses returned to North America as domestic animals with European conquest of the New World. Wild horse populations of the American West became established from domestic horses that escaped or were intentionally released.

Horses are large, grazing animals with long legs adapted for running. The feet are reduced to a single digit with a large, hard hoof which is the modified nail of the middle toe. This extreme adaptation extends the stride of the animal, enabling it to outrun predators. The cheek teeth of the horse are adapted for grinding and its long intestines for digestion of plant matter. All horses have large upper and lower incisors for cropping grass. Male horses have upper canine teeth, though females lack canines.

Animal Sign: Hoofprints that are roundish or oval ("horseshoe-shaped") with a V-shape in the middle; piles of brown scat roughly the size and shape of charcoal briquettes.

Wild Horse
Equus caballus

Field ID: This large mammal has a sturdy, muscular body; long legs with a one-toed hoof; long, boxy head; large eyes; upright ears; and a mane and long tail of coarse hair. It varies greatly in color and body pattern.

Size: Height at shoulder 4¼–4½ feet (130–137 cm), weight 660–1100 lbs (300–500 kg).

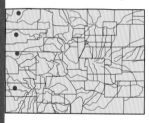

Habitat: Grasslands, mountain parks, rangeland, shrublands, piñon-juniper woodlands.

Distribution: Limited to four wild horse Herd Management Areas in western Colorado: Piceance Basin west of Meeker, Little Book Cliffs northeast of Grand Junction, Sandwash Basin northwest of Craig, and Spring Creek southwest of Montrose.

Field Notes: Wild horses, also known as mustangs, are actually feral domestic horses that became established in the wild. The state's four herds of wild horses are managed on public land by the Bureau of Land Management (BLM). They feed on native grasses and shrubs like winterfat and live in herds of females and their young led by a dominant stallion. Nonbreeding, subordinate males live in bachelor herds. Horses breed in spring and summer, and a single foal is born the next spring. Herd members communicate with neighs, whinnies, shrieks, and head and ear positions.

Because of the low quality of the forage on western rangelands, where they compete with wildlife and domestic livestock for food, wild horses are generally smaller and lighter than domestic horses. Mortality is usually due to starvation and the effects of harsh winter conditions. Wild horses are protected under the federal Wild and Free-roaming Horse and Burro Protection Act. To manage populations, BLM periodically rounds up excess horses, which are made available for adoption by individual owners.

Legal Status: Federally protected and managed.

Nonwild Wildlife

Two of Colorado's large, hoofed mammals linger in a twilight zone between being wildlife and domestic animals. The wild horse and bison, both iconic symbols of the American West, have essentially traveled opposite paths to their status today. The American bison, or buffalo, was once the most numerous large mammal in North America. Extirpated from the wild in Colorado, it is now a livestock animal raised on commercial ranches for its meat. The wild horse was brought to North America by Europeans as a domestic animal. Today it roams free in western Colorado. Neither is a wildlife species by Colorado statute. In a way, both bison and wild horses are nonwild wildlife—two living bits of history that still embody the wild character of Colorado.

ORDER
ARTIODACTYLA

Even-toed Hoofed Animals: Deer, Pronghorn, Bison, Goats, and Sheep

This diverse order includes most of the familiar large mammals of Colorado. They all have hooves and "headgear"—either antlers (grown only by males) or horns (grown by both males and females). All are ruminants with a complex digestive system that has multiple pouches to digest the plant cellulose they feed on. Food is partially digested then regurgitated into the mouth where it is further chewed and broken down, a process known as "chewing the cud." The cud is then swallowed again and continues moving through the digestive system. This not only allows ruminants to make use of an abundant food source (grass and other fibrous plants) but also lets them feed intensively, then retire to cover and continue digesting in a place safe from predators.

Artiodactyls have flexible spines and very long legs as an adaptation for running. They walk on the highly modified nails of the third and fourth toes, their "cloven hoof." This adaptation gives them great extension of the leg and foot and thus a very long stride. The first digit, comparable to a human's thumb or big toe, is absent, but all except pronghorn retain the second and fifth digits as vestigial toes, called dewclaws or dew hooves, which sometimes appear in the track.

The molars of these animals are adapted for crushing and chewing high-cellulose plants like grass. They lack upper incisors and canine teeth, and the jaws have a gap, called the diastema, between the cheek-teeth and incisors. The large eyes are on the side of the skull, allowing them to watch for predators while feeding and to have a wide field of vision. Most have white rumps and when frightened, flare these white hairs as they flee as an alarm signal to other animals.

Family Cervidae—Deer

Animals of the deer family are very recognizable, especially the males with antlers. They are all large and graceful (some might argue whether this description fits a moose!), with sturdy bodies, long legs, long necks, and wedge-shaped heads. Males have branched antlers most of the year, females are antler-less. They all have cloven hooves, with two vestigial dew hooves (also called dewclaws). Cervids are grazers or browsers.

Members of the deer family vary in their sociability. Cow elk, their calves, and sub-adults live together much of the year, and mule deer may gather in large herds in winter. White-tailed deer cluster in small groups. Moose are generally solitary, except for a cow with her calf.

Animal Sign: Dropped antlers; cloven hoof track, sometimes showing additional two smaller marks of dew hooves behind main hooves; shrubs and branches thrashed by males; rub trees with bark rubbed off by antlers; bark stripped from trees with cut mark (made by lower incisors) at the bottom and shredded rip mark at upper edge; ovals of matted grass beds.

Elk
Cervus elaphus

Family Cervidae—Deer

Field ID: This large deer is grayish tan with a white rump, dark brown neck and head, and dark legs. For much of the year, the males have large antlers.

Size: Head/body length 6²/₃–8¹/₃ feet (2–2.5 m), tail 4–8½ inches (102–216 mm), height at shoulder 4–5 feet (122–152 cm); weight, male 700–1000 lbs (318–454 kg), female 500–600 lbs (227–272 kg); antler height to 65 inches (165 cm), antler spread to 74 inches (188 cm).

Habitat: Mountain parks, mountain meadows, alpine tundra.

Distribution: Western two-thirds of the state, generally at elevations above 6,000 feet.

Field Notes: The North American elk, also called *wapiti*, is one of the most dramatic animals of the Colorado landscape. Elk are grazers, but they also browse on shrubs and the twigs and bark of aspen. In summer, they inhabit high elevation mountain parks, descending to lower elevations in winter. Through much of the year, elk live in separate herds of females (called cows) and their calves, and bachelor herds of males (bulls). In late September and October, they gather in open mountain parks for the fall rut. Bulls spend a great deal of energy rounding up females into harems. They challenge other males and attract females with shrill vocalizations called bugling. Competing males may spar antler-to-antler. Females leave the herd in early summer to bear their single calf in an isolated place, rejoining the herd when the calf is two to three weeks old. The antlers of mature bulls may be 3 or more feet long, consisting of a single main beam with up to six or more branches, or tines. Yearling males, called spike bulls, grow short, unbranched antlers.

Legal Status: Big game.

Mule Deer
Odocoileus hemionus

Field ID: These medium-sized deer are reddish tan to grayish brown, with pale rumps, pale muzzles with black foreheads, long legs, and large ears. The ropy tail is white with a black tip. Males have antlers much of the year.

Size: Head/body length 3²/₃–5 feet (112–152 cm), tail 4–8½ inches (102–216 mm), height at shoulder 3–3½ feet (91–107 cm); weight, male 125–400 lbs (57–181 kg), female 100–150 lbs (45–68 kg); antler spread to 47 inches (119 cm).

Habitat: Grasslands, shrublands, piñon-juniper woodlands, open mountain forests, riparian areas, alpine tundra.

Distribution: Statewide.

Field Notes: Named for their large, mulelike ears, "muleys" are the classic deer of the West. They inhabit nearly every habitat in the state. They browse on tender vegetation and also graze on grass. They can be distinguished from white-tailed deer by their long, black-tipped tails, large ears, and dark foreheads. The bucks' antlers branch into two main beams, each in turn splitting into two or more tines. When alarmed, they flee in a characteristic bounding gait called pronking, spronking, or stotting, in which all four legs leave the ground at the same time. Much of the year, bucks are solitary or live in groups of two to five males. During the November–December rut, the necks of the bucks swell significantly. In winter, mule deer gather in large numbers on favored wintering grounds. In spring, does go off alone to drop their fawns. First-year mothers usually have one fawn; older mothers typically bear twins. The mother and young gather with other does, fawns, and yearlings once the fawns are a few months old. Market hunting nearly eradicated mule deer from Colorado by the early twentieth century, but game management laws helped recover the species, and it is an important big game animal.

Legal Status: Big game.

Keeping Warm

Mammals are warm-blooded animals with an internal metabolic "furnace" that allows them to live in northern climates and high altitudes. But it is good insulation that keeps that internal heat from being lost to the environment.

Many animals, from foxes to mink to mountain goats, grow a double coat of hair to keep their bodies warm and dry. An outer coat of long, coarse guard hairs seals out moisture and seals in a layer of air around the body. An inner coat of thick, dense underfur that is wooly or curly creates this layer by trapping pockets of air in the spaces between the fibers. The body's heat then warms this air layer and allows the animal to keep warm enough to survive. Oil glands secrete water-proofing oil that seals the inner and outer coats against water and heat loss.

Some animals, such as deer, have hollow hair that serves the same function as the air spaces between hairs. Hollow hairs also make an animal more buoyant in water and help it swim.

White-tailed Deer
Odocoileus virginianus

Field ID: This medium-sized deer is grayish to reddish brown, with paler underparts and a broad tail with white undersides. Males have antlers much of the year.

Size: Head/body length 3¼–4 feet (107–122 cm), tail 5¾–13 inches (15–33 cm), height at shoulder 3–3½ feet (91–107 cm); weight, male 75–400 lbs (33–180 kg), female 50–250 lbs (23–113 kg); antler spread to 33 inches (84 cm).

Habitat: River and stream bottoms, riparian areas, croplands.

Distribution: Eastern Colorado from the foothills eastward, some mountain parks, San Luis Valley, and along the Colorado River.

Field Notes: The mule deer is a western species, but white-tailed deer are distributed through most of North America—from the Canadian prairie southward into Central America and northern South America. In Colorado, whitetails are animals mainly of Eastern Plains streams and rivers, where they shelter in riparian woodlands. More shy than mule deer, they bound away through the trees when disturbed, raising their broad, white tails as warning "flags." The antlers have a main, curving beam with two or more branches projecting from it, which may in turn branch into additional tines. Deer are primarily browsers, feeding on tender leaves, twigs, and shoots of shrubs and trees, though they will eat grass. Whitetails also feed on mushrooms, acorns, fruit, and crops such as wheat and corn. During the November rut, the necks of the bucks swell. Bucks rub their antlers on trees and mark their territories by urinating in characteristic scrapes pawed in the soil. Females give birth to one to three fawns in spring. In winter, they gather in herds composed of females and their offspring from several seasons.

Legal Status: Big game.

Moose
Alces americanus

Field ID: This largest member of the deer family is dark brown to nearly black with a paler belly. It has long legs, large ears, a long, bulbous muzzle, and a flap of skin, called a dewlap or bell, dangling from the throat. The shoulders are higher than the hips, and the short tail is inconspicuous. Males have very large, fanlike antlers that spread out and back from the head.

Size: Head/body length 7¾–9⅛ feet (2.4–2.8 m), tail 3–4½ inches (76–114 mm), height at shoulder 5–6½ feet (1.5–2 m); weight, male 830–1200 lbs (376–544 kg), female 600–800 lbs (272–363 kg); antler spread to 77 inches (196 cm).

Habitat: Mountain wetlands, marshes, streams and rivers, beaver ponds, wet forest edges.

Distribution: North and Middle parks, Rocky Mountain National Park, Grand Mesa; occasional sightings along waterways in the high mountains, in the eastern foothills, and in suburban areas.

Field Notes: Moose are solitary animals that inhabit mountain wetlands and marshy meadows and riverways. Their long legs and broad hooves allow them to wade into marshy habitat to feed on semiaquatic vegetation. They can submerge for several minutes. Moose may consume as much as 25 pounds of vegetation daily. They are solitary except for cows with calves, though in winter they may gather in areas with good habitat. This isn't a social

Family Cervidae—Deer

grouping; the feeding animals seemingly ignore each other. Instead of gathering for a dramatic rut and collecting harems of females, bull moose locate individual receptive females. Cow moose bear a single calf, occasionally twins, in late May or early June. They can be aggressive when surprised or disturbed. Moose were never abundant in Colorado, which is at the southern edge of the species' range, and may not have regularly inhabited the state prior to European settlement. Beginning in 1978, Colorado Parks and Wildlife (then known as the Division of Wildlife) released moose in several areas of the state as a game species. Moose are now well established, though populations are very localized and not large. The Moose Center in the Colorado State Forest near Walden is an excellent place to learn about moose.

Legal Status: Big game.

Family Antilocapridae—Pronghorn

The pronghorn is the sole surviving species of this once-diverse family, which flourished in North America tens of millions of years ago during the Age of Mammals. Pronghorn are not closely related to Old World antelope, but when American explorer Meriwether Lewis (who made the first European record of the species) first saw these exotic-looking creatures, he named them for what he thought they resembled. The family name Antilocapridae translates as "antelope-goat."

The pronghorn is an animal of open western prairies, shrub-lands and rangelands. Its brown-and-white, horizontal counter-shading makes it very hard to spot at a distance. The pronghorn is highly adapted for escaping predators. Not only is it exceptionally fast, it is also extremely wary and flees at even a distant approach by potential predators. Bucks often stand watch over a feeding group and will snort and stamp as a warning. The pronghorn's eyes are located high on the sides of the head, enabling the animal to keep watch even while its head is down feeding. Its very slender, long legs are more fully adapted for running than other artiodactyls, having lost the vestigial dew hooves that deer, bison, sheep, and goats retain.

Animal Sign: Slender, cloven hoof track in open country; piles of dark pellets; large, pale shapes visible on open prairies or range-land; shed horns.

Pronghorn
Antilocapra americana

Field ID: This deer-sized animal is reddish brown to tan with white sides, underparts and rump. It has a large head, ears, and eyes, and black horns that curve into a prong at the tip. There are two white bands across the throat.

Size: Head/body length 3¼–4½ feet (99–130 cm), tail 3–7 inches (76–178 mm), height at shoulder 3 feet (91 cm); weight 74–130 lbs (34–59 kg).

Habitat: Grasslands, mountain parks, semi-desert shrublands.

Distribution: Eastern Colorado, mountain parks and valleys of central Colorado, northwestern Colorado, San Luis Valley.

Field Notes: Mistakenly called antelope, pronghorn are not related to Old World antelope. They are open-country dwellers and built for speed. Pronghorn have been clocked at up to 60 miles per hour. They evolved their speed during the Age of Mammals, when cheetahs, the fastest land mammals, inhabited North America. Pronghorn are extremely wary and will flee at the slightest approach, flaring the white hairs on their rump as an alarm signal. Their vision is incredibly keen, and they avoid wooded or other habitats with an obscured field of vision. Pronghorn browse on sagebrush, bitterbrush, shrubs, forbs, and some grass. Both males and females have horns, though the females' are short, unbranched stubs. The horns are made of modified, fused hairs over a bony core but are shed annually like antlers. In late summer, bucks gather does into small harems. Twin fawns are born the next year in early summer. Within minutes of birth, fawns are able to walk on their own, though the doe conceals her newborn young in the grass while she feeds, returning to nurse them every few hours. Within about two weeks, the fawns can run well enough to elude predators, and the mother and her babies join other mothers and young in nursery herds. Much of the year, males live alone or join with other males in bachelor herds of ten or more animals. During winter, large groups combining all sexes and ages may gather on good rangeland.

Legal Status: Big game.

Antlers or Horns?

The dramatic structures adorning the heads of large, hoofed mammals define these animals. But there are major differences between an antler and a horn. Mountain goats, bighorn sheep, and bison have horns. Horns are made of keratin, a protein material similar to that of fingernails. Horns grow over a bony core that projects from the skull. Horns remain permanently on the animal, growing some each year. Both male and female goats, sheep, and bison have horns, but those of males are larger and more elaborate.

Antlers are the headgear of deer, elk, and moose. They are basically modified bone, made of crystalline minerals such as calcium, and grow from projections on the skull known as antler pads. Antlers are shed each year, and only male deer grow antlers.

The horns of pronghorn are sort of in between horns and antlers. The horns are sheaths made of modified hair that grow over a bony core, but they are shed each year. Both male and female pronghorn grow horns.

The amount of antler or horn grown in a season, and sometimes the shape, can vary depending on the quality and amount of food, the age of the animal, and its overall health. An aging bull elk, unable to eat sufficient nutrients, may grow deformed antlers. An antler or horn injured while growing (before it hardens) may also be misshapen.

Family Antilocapridae—Pronghorn

245

Family Bovidae—Cattle, Goats, and Sheep

This large and diverse family includes the wild cousins of familiar domestic animals—cows, sheep, and goats—as well as African and Asian antelope.

North American bovids have sturdy bodies, long legs, cloven hooves, and horns grown by both males and females. There are two dew hooves, which are vestigial toes, behind the hoof. Bovids are grazers, with cheek teeth modified for crushing and chewing vegetation.

Bovids are variously adapted for their lifestyle. Sheep and goats climb and bound easily in steep, rocky terrain. Bison inhabit open grasslands. When deep snow covers their forage, they swing their heavy heads to clear the snow, something domestic cattle can't do.

Bison live year-round in large herds of both sexes and all ages. Most of the year, male goats and sheep live in bachelor herds separate from females and their young, with both sexes and all ages gathering together during the rut.

Sheep and goats are quite wary and watchful and will flee at the approach of potential predators, generally fleeing upslope, away from threats approaching from below. Bison are less reactive to would-be predators but may charge when threatened. They typically form a defensive cluster, backed tightly together with young on the interior of the group, then females, then males on the outer edge, facing the threat.

Animal Sign: Cloven hoof tracks—goat and sheep are triangular, bison are round; piles of pellets or cowlike dung that dries in a patty.

Bison
Bison bison

Field ID: This very large, heavy-bodied bovine has a large, rounded head; long, ropy tail with a tuft at the end; and short, curving horns. The body is brown, with a darker, wooly mane over head and shoulders. It is taller at the shoulder than hip.

Size: Total length 5¼–10½ feet (1.58–3.2 m), tail 15½–23½ inches (39–60 cm), height at shoulder 5–6 feet (1.5–1.8 m); weight, male 1,012–2,860 lbs (460–1300 kg), female 890–1196 lbs (403–543 kg); horn spread to 35 inches (89 cm).

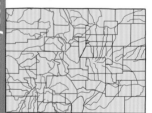

Habitat: Grasslands, mountain parks, western valleys, piñon-juniper woodlands, arid shrublands, alpine tundra, commercial ranches.

Distribution: Extirpated in the wild but formerly statewide.

Field Notes: The American bison, or buffalo, is the iconic animal of the American West. Millions once inhabited the Great Plains, moving in vast herds, their grazing, trampling and wallowing a major factor in the ecology of the prairies. Within a few decades in the mid-to-late nineteenth century, settlers and commercial hunters killed millions of wild bison and destroyed the great herds, both for meat and hides and to destroy the economic base of Plains Indians. By the beginning of the twentieth century, only about 500 wild bison survived from the herds of millions, mostly in Yellowstone National Park. Bison were at one time found throughout Colorado, with the possible exception of the San Juan Mountains and the Uncompahgre Plateau. The last wild bison in Colorado were killed between 1897 and 1904. The species was designated as livestock in 1975, and all bison today are managed either on commercial ranches or in public herds like those owned by the City of Denver. These large, wild cattle live in large herds of mixed ages. Bulls are significantly larger than cows. They mate in late summer to early fall and a single, pale chestnut-colored calf is born in the spring. Bison are grazers, each animal consuming 16 or more pounds of grass per day. The Denver Mountain Parks herd is easily visible along Interstate 70 at Genesee Park, exit 254. There is a second herd at Daniels Park in Douglas County.

Legal Status: Livestock.

Mountain Goat
Oreamnos americanus

Field ID: These sturdy wild goats have long white coats, white "beards," longish, angular faces, black hooves, and small, back-curving black horns.

Size: Total length 2¾–5 feet (84–152 cm), tail 3–8 inches (76–203 mm), height at shoulder 3–3½ feet (91–107 cm), weight 88–300 lbs (40–136 kg), horn spread to 11½ inches (29 cm).

Habitat: Cliffs, talus slopes, steep mountainsides at or above timberline, alpine tundra.

Distribution: Mount Evans, Gore Range, Collegiate Peaks, San Juan Mountains.

Field Notes: Goats are denizens of high, inaccessible cliffs and mountain slopes, where they graze on native grasses and forbs. When threatened, they climb upward, using topography as a defense against predators. Their hooves have spongy pads that help them grip on steep cliffs and slopes. Hikers may find tufts of long, white goat hair snagged on shrubs and rocks on high mountain slopes. Despite the austerity of their high mountain habitat, mountain goats may spend winter on windswept ridges. During severe cold and snow, they migrate down to timberline, sheltering in the trees. The goat's long coat of hollow hairs over a thick woolly undercoat provides excellent insulation. Goats do not move far from their traditional home range, thus goat populations are very specific in location and are readily found. Nannies and kids live in herds separate from billies until fall. Goats mate during the November–December rut, when males threaten each other by posing and showing their horns, though they rarely actually fight. One or two kids are born in spring. There is no evidence that mountain goats were native to Colorado prior to their introduction as a game species between the 1940s and 1970s.

Legal Status: Big game.

Rocky Mountain Bighorn Sheep
Ovis canadensis canadensis

Field ID: These sturdy mountain sheep are grayish to brown with cream to white rumps and distinctive, heavy horns that are gently curved on ewes and curled on mature rams.

Size: Head/body length 2¼–3 feet (69–91 cm), tail 2¾–5 inches (70–127 mm), height at shoulder 2½–3½ feet (76–107 cm); weight, male 125–272 lbs (57–123 kg), female 75–150 lbs (34–68 kg); horn spread to 33 inches (84 cm).

Habitat: Mountain slopes, cliffs.

Distribution: Through the central mountains and foothills, with disjunct populations in Douglas County and the canyons of southeastern Colorado.

Field Notes: These mountain sheep—the state mammal and official symbol of Colorado Parks and Wildlife—embody the wild character of Colorado. They are named for the massive horns of the rams, which grow over successive seasons into a full curl that flares out from the head. The horns of ewes are smaller and thinner but also curve back to no more than a half-curl. Though associated now with rugged mountain landscapes, bighorn were once much more broadly distributed, even inhabiting grasslands along the foothills. Sheep graze on grass on steep, open, often south-facing hillsides. They prefer less steep terrain than mountain goats, often at lower elevations. They are more likely to outrun a threat than to climb to avoid it, though their flexible feet have spongy pads to help cling to rocks. Much of the year, ewes and lambs live in nursery herds and males gather in bachelor herds. During the December rut, all ages and sexes of sheep gather in open mountain meadows where the males compete with each other by charging and colliding horn-to-horn. The resulting crash can be heard a mile away. Two rams may battle until one staggers away, and the other mates with the assembled ewes. Ewes bear a single lamb in early summer on lambing grounds that are used year after year. The lambs are able to follow their mothers very soon after birth.

Legal Status: Big game.

Desert Bighorn Sheep
Ovis canadensis nelsoni

Field ID: These mountain sheep are sandy brown with cream to white rumps. Their horns are more slender and flatter than Rocky Mountain bighorn and spread out more from their heads.

Size: Head/body length 2¼–3 feet (69–91 cm), tail 2¾–5 inches (70–127 mm), height at shoulder 2½–3½ feet (76–107 cm); weight, male 125–200 lbs (57–91 kg), female 77–150 lbs (35–68 kg); horn spread to 33 inches (84 cm).

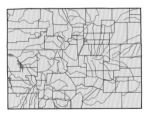

Habitat: Canyons, cliffs, rocky hillsides.

Distribution: Western Colorado, west and south of Grand Junction.

Field Notes: Desert bighorn are a subspecies of *Ovis canadensis*, more slender, less stocky, and paler than the native Rocky Mountain bighorn of the high mountains. Colorado Parks and Wildlife introduced desert bighorn as a game animal into the canyon lands of western Colorado, near Colorado National Monument, in 1979. Desert bighorn are now well established in these canyons. They feed on native grasses and browse on desert shrubs. Their horns are more slender, flatter and wider-spreading than those of Rocky Mountain bighorn, though they grow into a similar curl around and out from the head. Desert bighorn don't require drinking water in winter if sufficient succulent or green vegetation is available. During the hot summer months, when their bodies use more water, their movements revolve around water sources, and they may reduce activity to conserve moisture by resting during the day beneath shadowed overhangs and sheltered ledges. The rut, which occurs earlier than it does for Rocky Mountain bighorn, may begin in July. Hikers and backpackers have a good chance of seeing desert bighorn in Dominguez Canyon, southwest of Grand Junction.

Legal Status: Big game.

Where to See Colorado Mammals

Mammals are all around us, but because many of them are small, secretive, and nocturnal, we rarely see them. But large mammals like elk and bighorn sheep have predictable habits and habitats. Here are some of the best spots in the state to see wildlife:

- Rocky Mountain National Park—In fall, this is the premier spot to witness elk bugling. Bighorn sheep, pikas, and marmots are also often easily visible along Trail Ridge Road in the summer. Pine and Abert's squirrels inhabit forests, and beavers inhabit the many open creeks. Moose are often spotted on the west side of the Park. Ground squirrels and chipmunks are common.
- Mount Evans—Drive the Mount Evans Highway for a good chance of spotting bighorn sheep and mountain goats as well as elk, mule deer, marmots, and pikas.
- Georgetown State Wildlife Area—Great views in winter of bighorn sheep grazing on the south-facing slope across Interstate 70 from the viewing platform.
- Arapaho National Wildlife Refuge, North Park—Moose, pronghorn, mule deer, coyotes, badgers, prairie dogs, jackrabbits, ground squirrels, and many other mammals can be seen here.
- Pawnee National Grassland—Pronghorn, prairie dogs, ground squirrels, coyotes, swift foxes, jackrabbits, and other prairie species inhabit this remnant of prairie in northeastern Colorado.
- South Park—This large, wide-open mountain valley is home to pronghorn, mule deer, elk, coyotes, foxes, numerous small mammals, and a herd of ranched bison near Hartsel.
- San Luis Valley—Rangelands throughout the valley are home to pronghorn and a wide variety of small mammals. Look for bison on Medano-Zapata Ranch, just west of Great Sand Dunes National Park and Preserve.

The *Colorado Wildlife Viewing Guide*, by Mary Taylor Young, details 201 sites throughout the state to see wildlife.

Checklist of
Colorado Mammals

Scientific names used in this guide are based upon Wilson and Reeder's *Mammal Species of the World, Third Edition*, 2005. Common names mostly follow this source, with some variation for more familiar local names.

Order Didelphimorphia—American Marsupials
FAMILY DIDELPHIDAE—OPOSSUMS
- ❏ Virginia Opossum, *Didelphis virginiana*

Order Soricomorpha—Shrews and Moles
FAMILY SORICIDAE—SHREWS
- ❏ Masked Shrew, *Sorex cinereus*
- ❏ Pygmy Shrew, *Sorex hoyi*
- ❏ Merriam's Shrew, *Sorex merriami*
- ❏ Montane Shrew, *Sorex monticolus*
- ❏ Dwarf Shrew, *Sorex nanus*
- ❏ American Water Shrew, *Sorex palustris*
- ❏ Elliot's Short-tailed Shrew, *Blarina hylophaga*
- ❏ Least Shrew, *Cryptotis parva*
- ❏ Desert Shrew, *Notiosorex crawfordi*

FAMILY TALPIDAE—MOLES
- ❏ Eastern Mole, *Scalopus aquaticus*

Order Chiroptera—Bats
FAMILY VESPERTILIONIDAE
- ❏ California Myotis, *Myotis californicus*
- ❏ Western Small-footed Myotis, *Myotis ciliolabrum*
- ❏ Long-eared Myotis, *Myotis evotis*
- ❏ Little Brown Bat/Little Brown Myotis, *Myotis lucifugus*
- ❏ Fringed Myotis, *Myotis thysanodes*
- ❏ Long-legged Myotis, *Myotis volans*
- ❏ Yuma Myotis, *Myotis yumanensis*
- ❏ Eastern Red Bat, *Lasiurus borealis*
- ❏ Hoary Bat, *Lasiurus cinereus*
- ❏ Silver-haired Bat, *Lasionycteris noctivagans*
- ❏ Canyon Bat/Western Pipistrelle, *Pipistrellus hesperus*
- ❏ Big Brown Bat, *Eptesicus fuscus*

- ❐ Spotted Bat, *Euderma maculatum*
- ❐ Townsend's Big-eared Bat, *Corynorhinus townsendii*
- ❐ Pallid Bat, *Antrozous pallidus*

FAMILY MOLOSSIDAE—FREE-TAILED BATS

- ❐ Brazilian Free-tailed Bat, *Tadarida brasiliensis*
- ❐ Big Free-tailed Bat, *Nyctinomops macrotis*

Order Cingulata—Armadillos

FAMILY DASYPODIDAE

- ❐ Nine-banded Armadillo, *Dasypus novemcinctus*

Order Lagomorpha—Rabbits and Relatives

FAMILY OCHOTONIDAE—PIKAS

- ❐ American Pika, *Ochotona princeps*

FAMILY LEPORIDAE—RABBITS AND HARES

- ❐ Desert Cottontail, *Sylvilagus audubonii*
- ❐ Eastern Cottontail, *Sylvilagus floridanus*
- ❐ Mountain Cottontail/Nuttall's Cottontail, *Sylvilagus nuttallii*
- ❐ Snowshoe Hare, *Lepus americanus*
- ❐ Black-tailed Jackrabbit, *Lepus californicus*
- ❐ White-tailed Jackrabbit, *Lepus townsendii*

Order Rodentia—Rodents

FAMILY SCIURIDAE—SQUIRRELS

- ❐ Cliff Chipmunk, *Tamias dorsalis*
- ❐ Least Chipmunk, *Tamias minimus*
- ❐ Colorado Chipmunk, *Tamias quadrivittatus*
- ❐ Hopi Chipmunk, *Tamias rufus*
- ❐ Uinta Chipmunk, *Tamias umbrinus*
- ❐ Yellow-bellied Marmot, *Marmota flaviventris*
- ❐ White-tailed Antelope Squirrel, *Ammospermophilus leucurus*
- ❐ Wyoming Ground Squirrel, *Spermophilus elegans*
- ❐ Golden-mantled Ground Squirrel, *Spermophilus lateralis*
- ❐ Spotted Ground Squirrel, *Spermophilus spilosoma*
- ❐ Thirteen-lined Ground Squirrel, *Spermophilus tridecemlineatus*
- ❐ Rock Squirrel, *Spermophilus variegatus*
- ❐ Gunnison's Prairie Dog, *Cynomys gunnisoni*
- ❐ White-tailed Prairie Dog, *Cynomys leucurus*
- ❐ Black-tailed Prairie Dog, *Cynomys ludovicianus*
- ❐ Abert's Squirrel, *Sciurus aberti*
- ❐ Fox Squirrel, *Sciurus niger*

❑ Pine Squirrel (Chickaree), *Tamiasciurus hudsonicus*

FAMILY GEOMYIDAE—POCKET GOPHERS

❑ Botta's Pocket Gopher, *Thomomys bottae*

❑ Northern Pocket Gopher, *Thomomys talpoides*

❑ Plains Pocket Gopher, *Geomys bursarius*

❑ Yellow-faced Pocket Gopher, *Cratogeomys castanops*

FAMILY HETEROMYIDAE—POCKET MICE AND KANGAROO RATS

❑ Olive-backed Pocket Mouse, *Perognathus fasciatus*

❑ Plains Pocket Mouse, *Perognathus flavescens*

❑ Silky Pocket Mouse, *Perognathus flavus*

❑ Great Basin Pocket Mouse, *Perognathus parvus*

❑ Hispid Pocket Mouse, *Chaetodipus hispidus*

❑ Ord's Kangaroo Rat, *Dipodomys ordii*

FAMILY CASTORIDAE—BEAVERS

❑ American Beaver, *Castor canadensis*

FAMILY CRICETIDAE—NATIVE MICE, RATS, AND VOLES

❑ Western Harvest Mouse, *Reithrodontomys megalotis*

❑ Plains Harvest Mouse, *Reithrodontomys montanus*

❑ Brush Mouse, *Peromyscus boylii*

❑ Canyon Mouse, *Peromyscus crinitus*

❑ White-footed Mouse, *Peromyscus leucopus*

❑ Deer Mouse, *Peromyscus maniculatus*

❑ Northern Rock Mouse, *Peromyscus nasutus*

❑ Piñon Mouse, *Peromyscus truei*

❑ Northern Grasshopper Mouse, *Onychomys leucogaster*

❑ Hispid Cotton Rat, *Sigmodon hispidus*

❑ White-throated Woodrat, *Neotoma albigula*

❑ Bushy-tailed Woodrat, *Neotoma cinerea*

❑ Eastern Woodrat, *Neotoma floridana*

❑ Desert Woodrat, *Neotoma lepida*

❑ Mexican Woodrat, *Neotoma mexicana*

❑ Southern Plains Woodrat, *Neotoma micropus*

❑ Southern Red-backed Vole, *Myodes gapperi*

❑ Western Heather Vole, *Phenacomys intermedius*

❑ Long-tailed Vole, *Microtus longicaudus*

❑ Mexican Vole, *Microtus mexicanus*

❑ Montane Vole, *Microtus montanus*

❑ Prairie Vole, *Microtus ochrogaster*

❑ Meadow Vole, *Microtus pennsylvanicus*

❑ Sagebrush Vole, *Lemmiscus curtatus*

- ☐ Muskrat, *Ondatra zibethicus*

FAMILY MURIDAE—OLD WORLD MICE AND RATS

- ☐ House Mouse, *Mus musculus*
- ☐ Norway Rat, *Rattus norvegicus*

FAMILY DIPODIDAE—JUMPING MICE AND JERBOAS

- ☐ Meadow Jumping Mouse, *Zapus hudsonius*
- ☐ Western Jumping Mouse, *Zapus princeps*

FAMILY ERETHIZONTIDAE—NEW WORLD PORCUPINES

- ☐ North American Porcupine, *Erethizon dorsatum*

Order Carnivora—Carnivores

FAMILY CANIDAE—DOGS

- ☐ Coyote, *Canis latrans*
- ☐ Gray Wolf, *Canis lupus*
- ☐ Kit Fox, *Vulpes macrotis*
- ☐ Swift Fox, *Vulpes velox*
- ☐ Red Fox, *Vulpes vulpes*
- ☐ Gray Fox, *Urocyon cinereoargenteus*

FAMILY URSIDAE—BEARS

- ☐ Black Bear, *Ursus americanus*
- ☐ Grizzly Bear, *Ursus arctos*

FAMILY PROCYONIDAE—RACCOONS AND ALLIES

- ☐ Ringtail, *Bassariscus astutus*
- ☐ Raccoon, *Procyon lotor*

FAMILY MUSTELIDAE—WEASELS

- ☐ Pine (American) Marten, *Martes americana*
- ☐ Short-tailed Weasel (Ermine), *Mustela erminea*
- ☐ Long-tailed Weasel, *Mustela frenata*
- ☐ Black-footed Ferret, *Mustela nigripes*
- ☐ Mink, *Neovison vison*
- ☐ Wolverine, *Gulo gulo*
- ☐ American Badger, *Taxidea taxus*
- ☐ Northern River Otter, *Lontra canadensis*

FAMILY MEPHITIDAE—SKUNKS

- ☐ Western Spotted Skunk, *Spilogale gracilis*
- ☐ Eastern Spotted Skunk, *Spilogale putorius*
- ☐ Striped Skunk, *Mephitis mephitis*

FAMILY FELIDAE—CATS

- ☐ Mountain Lion, *Puma concolor*
- ☐ Canada Lynx, *Lynx canadensis*

❏ Bobcat, *Lynx rufus*

Order Perissodactyla—Odd-toed Hoofed Animals

FAMILY EQUIDAE—HORSES

❏ Wild Horse, *Equus caballus*

Order Artiodactyla—Even-toed Hoofed Animals

FAMILY CERVIDAE—DEER

❏ Elk, *Cervus elaphus*

❏ Mule Deer, *Odocoileus hemionus*

❏ White-tailed Deer, *Odocoileus virginianus*

❏ Moose, *Alces americanus*

FAMILY ANTILOCAPRIDAE—PRONGHORN

❏ Pronghorn, *Antilocapra americana*

FAMILY BOVIDAE—CATTLE, GOATS, AND SHEEP

❏ Bison, *Bison bison*

❏ Mountain Goat, *Oreamnos americanus*

❏ Rocky Mountain Bighorn Sheep, *Ovis canadensis canadensis*

❏ Desert Bighorn Sheep, *Ovis canadensis nelsoni*

Mammals with Limited Colorado Range

These species may inhabit Colorado, but there are only limited records of them in the state or they are likely present but have not yet been documented.

EASTERN PIPISTRELLE/TRI-COLORED BAT—*PERIMYOTIS SUBFLAVUS*

A single adult female was collected on the side of a dwelling in Greeley in 1987.

ALLEN'S BIG-EARED BAT—*IDIONYCTERIS PHYLLOTIS*

A single confirmed record from calls recorded in La Sal Creek canyon in 2006.

NORTHERN FLYING SQUIRREL—*GLAUCOMYS SABRINUS*

Not known to inhabit Colorado but confirmed in southern Wyoming and extreme northeastern Utah.

Whose Skull Is This?

When you find a skull in the field, you can get some clues to its identity by examining its size, shape, and structure. Where are the eyes located? Are they on the side of the head, like those of a rabbit or other plant-eater that needs to keep a wary watch for predators? Are they in front, like those of a cat, which uses binocular vision to judge the distance to a prey animal? Imagine what the skull would look like covered with muscle and fur to give further thoughts on what animal it might be. If it is a small skull, is it that of a mouse or a larger rodent, like a woodrat?

One of your best identification clues is the teeth. Biologists use teeth as an important characteristic for identifying mammals. A species' teeth are adapted for its particular lifestyle—shearing teeth for meat-eaters, grinding teeth for grazers, chiseling teeth for gnawing animals.

Carnivores have pointed canine teeth. Many plant-eaters lack canine teeth. A tiny skull with a long muzzle that has canine teeth may be a shrew. A small skull with a long muzzle that lacks canines but has long, brown or orange-fronted incisors may be a mouse.

The number of each kind of tooth, known as the dental formula, helps identify a mammal, pinpointing its genus and sometimes even its species. The following table of dental formulas will help identify skulls found in the field or in scat or owl pellets.

Dental Formulas

Each formula represents the teeth on one side of the skull (the other side is a mirror image with the same count). The top number is for upper teeth (U), the bottom number for lower jaw teeth (L). Formulas are expressed in the sequence of kinds of teeth from front of mouth to back: incisors (I), canines (C), premolars (P), molars (M). A zero indicates that no teeth of that particular kind are present, usually leaving a space along the jaw bone.

Species	UI/LI	UC/LC	UP/LP	UM/LM	Total Teeth
Virginia opossum	5/4	1/1	3/3	4/4	50
coyote, gray wolf, kit fox, swift fox, red fox, gray fox, black bear, grizzly bear	3/3	1/1	4/4	2/3	42
ringtail, raccoon	3/3	1/1	4/4	2/2	40
wild horse - male	3/3	1/1	3-4/3	3/3	40-42
wild horse - female	3/3	0/0	3-4/3	3/3	36-38
pine marten, wolverine	3/3	1/1	4/4	1/2	38
California myotis, western small-footed myotis, long-eared myotis, little brown bat, fringed myotis, long-legged myotis, Yuma myotis	2/3	1/1	3/3	3/3	38
river otter	3/3	1/1	4/3	1/2	36
Townsend's big-eared bat, silver-haired bat	2/3	1/1	2/3	3/3	36
eastern mole	3/2	1/0	3/3	3/3	36
canyon bat, spotted bat	2/3	1/1	2/2	3/3	34
short-tailed weasel, long-tailed weasel, black-footed ferret, mink, badger, western spotted skunk, eastern spotted skunk, striped skunk	3/3	1/1	3/3	1/2	34
elk	0/3	1/1	3/3	3/3	34
big brown bat	2/3	1/1	1/2	3/3	32
red bat, hoary bat, Brazilian free-tailed bat	1/3	1/1	2/2	3/3	32

Species	UI/LI	UC/LC	UP/LP	UM/LM	Total Teeth
masked shrew, pygmy shrew, Merriam's shrew, montane shrew, dwarf shrew, American water shrew, Elliot's short-tailed shrew	3/1	1/1	3/1	3/3	32
mule deer, white-tailed deer, moose, pronghorn, mountain goat, bison, bighorn sheep	0/3	0/1	3/3	3/3	32
armadillo	0/0	0/0		{8/8}	32
big free-tailed bat	1/2	1/1	2/2	3/3	30
least shrew	3/1	1/1	2/1	3/3	30
mountain lion	3/3	1/1	3/2	1/1	30
lynx, bobcat	3/3	1/1	2/2	1/1	28
pallid bat	1/2	1/1	1/2	3/3	28
desert shrew	3/1	1/1	1/1	3/3	28
desert cottontail, eastern cottontail, mountain cottontail, snowshoe hare, white-tailed jackrabbit, black-tailed jackrabbit	2/1	0/0	3/2	3/3	28
pika	2/1	0/0	3/2	2/3	26
cliff chipmunk, least chipmunk, Colorado chipmunk, Hopi chipmunk, Uinta chipmunk, yellow-bellied marmot, white-tailed antelope squirrel, rock squirrel, Gunnison's prairie dog, white-tailed prairie dog, black-tailed prairie dog, Abert's squirrel, Wyoming ground squirrel, golden-mantled ground squirrel, spotted ground squirrel, thirteen-lined ground squirrel	1/1	0/0	2/1	3/3	22

Species	UI/LI	UC/LC	UP/LP	UM/LM	Total Teeth
fox squirrel, pine squirrel, Botta's pocket gopher, northern pocket gopher, plains pocket gopher, yellow-faced pocket gopher, olive-backed pocket mouse, plains pocket mouse, silky pocket mouse, Great Basin pocket mouse, hispid pocket mouse, beaver, Ord's kangaroo rat, porcupine	1/1	0/0	1/1	3/3	20
meadow jumping mouse, western jumping mouse	1/1	0/0	1/0	3/3	18
western harvest mouse, plains harvest mouse, brush mouse, canyon mouse, white-footed mouse, deer mouse, northern rock mouse, piñon mouse, northern grasshopper mouse, hispid cotton rat, white-throated woodrat, bushy-tailed woodrat, eastern woodrat, desert woodrat, Mexican woodrat, Southern Plains woodrat, house mouse, Norway rat, southern red-backed vole, western heather vole, long-tailed vole, Mexican vole, montane vole, prairie vole, meadow vole, sagebrush vole, muskrat	1/1	0/0	0/0	3/3	16

Glossary

Billy—Male mountain goat

Bull—Male elk, moose, or bison

Browser—An animal that feeds on leaves, flowers, and soft vegetation other than grass; often connotes feeding on woody vegetation

Buck—Male deer or pronghorn

Calf—Young or baby bison, elk, or moose

Carnivore—An animal that eats mainly or exclusively animal flesh; sometimes used as a term for members of the order Carnivora

Carrion—The flesh of a dead animal that is not freshly killed by a predator

Castings—The soil excavated by fossorial animals

Coprophagy—The ingestion of feces to extract further nutrients

Cow—Female elk, moose, or bison

Crepuscular—Active during dawn and dusk

Cud—The pulpy, partly digested food regurgitated into the mouth by ruminants (deer, bison, pronghorn, goats, and sheep) to be re-chewed as part of the digestive process

Digitigrade—Walking on the toes of the foot, like coyotes and bobcats

Diurnal—Active during the day

Doe—Female deer or pronghorn

Dormancy—A state of reduced activity in response to harsh environmental conditions but not characterized by a dramatic reduction of metabolic processes

Echolocation—Identifying the location and shape of objects or prey by reading the echoes or "sound shadows" of ultrasonic vocalizations that bounce back from objects. Used by bats and shrews.

Endangered—When a species population has become so small that it is in imminent danger of extirpation or extinction

Eskers—Ridges of soil snaking across the ground that are exposed after snow melts; formed from excavated soil pushed to the surface by gophers and packed into tunnels in the snow

Estivation (aestivation)—A state of reduced metabolism, comparable to hibernation, but in response to extreme hot weather

Ewe—Female bighorn sheep

Extirpated—When no more members of a species exist in an area; locally extinct

Fawn—Young or baby deer or pronghorn

Forb—Nonwoody, nongrass flowering plant (a contraction of "forest herb")

Fossorial—Adapted to digging and living underground

Grazer—An animal that feeds on grass

Herbivore—An animal that eats mainly or exclusively vegetation

Hibernation—A state of reduced metabolism in response to winter cold. Respiration, heart rate, body temperature, and digestion are all greatly reduced and the animal does not eat, drink, urinate, or defecate.

Hispid—Covered with stiff or coarse hairs; bristly

Juvenile—A developmental stage of young animals, beyond newborn but prior to becoming sub-adults (adolescents); a generally used term lacking strict definition

Kid—Young or baby mountain goat

Nanny—Female mountain goat

Nocturnal—Active at night

Omnivore—An animal that eats both plant and animal matter

Plantigrade—Walking on the soles of the foot, like bears, raccoons, and people

Ram—Male bighorn sheep

Ruminant—An animal with a multipouched gut (esophagus and stomach) that regurgitates and rechews (ruminates) partially digested food as part of the process of digesting a high-cellulose plant diet

Sub-adult—A loose term for a developmental stage between young or juvenile and adult; adolescent

Threatened—When a species population has become so small that it is at risk of becoming endangered

Torpor—A state of reduced metabolism in response to stress in the environment (temperature, drought, etc.). The heart rate slows and the body temperature drops, but the animal is able to wake up and move around. Torpor is shorter and less deep than hibernation.

Tragus—A fleshy projection at the entrance of the ear of a bat that functions in triangulating the vertical position of prey from sound echoes received during echolocation

Ungulate—A hoofed mammal, including perissodactyls (horses and their kin) and artiodactyls, such as bison, deer, pronghorn, goats, and sheep

Unguligrade—Walking on a hoof, the extremely modified nail of one or more toes (see Ungulate)

Vestigial—Nonfunctioning organs or body parts that are still present in an animal's anatomy, though they may be greatly reduced

Other Resources

Adams, Rick. *Bats of the Rocky Mountain West: Natural History, Ecology and Conservation*. Boulder: University Press of Colorado, 2004.

Armstrong, David M. *Rocky Mountain Mammals: A Handbook of Mammals of Rocky Mountain National Park and Vicinity,* 3rd ed., Boulder: University Press of Colorado, 2008.

Armstrong, David M., James P. Fitzgerald, and Carron A. Meaney. *Mammals of Colorado,* 2nd ed. Denver: Denver Museum of Nature & Science; Boulder: University Press of Colorado, 2011.

Bat Conservation International
www.batcon.org

Colorado Parks and Wildlife website
http://wildlife.state.co.us

Kays, Roland W., and Don E. Wilson. *Mammals of North America.* Princeton, NJ: Princeton University Press, 2002.

Reid, Fiona. *A Field Guide to Mammals of North America,* 4th ed. New York: Houghton Mifflin Harcourt, 2006.

About the Author

Award-winning nature writer Mary Taylor Young spent her childhood summers roaming the Colorado Rockies from her grandparents' cabin in Estes Park. Her love of wild things and the outdoors led to a degree in zoology from Colorado State University and a life devoted to nature and the environment. She has written 13 books, including *The Guide to Colorado Reptiles and Amphibians*, hundreds of magazine and newspaper articles, and works extensively with the US Fish and Wildlife Service, US Forest Service, Colorado Parks and Wildlife, and Utah Division of Wildlife Resources. Young lives in Castle Rock, Colorado, with her husband, daughter, two dogs, and many wild neighbors.

Index

271